Genetics:
Science, Ethics,
and Public Policy

READINGS IN BIOETHICS

Edited by Thomas A. Shannon

Readings in Bioethics is a series of anthologies that gather together seminal essays in four areas: reproductive technologies, genetic technologies, death and dying, and health care policy. Each of the readers addresses specific case studies and issues under its respective topic. The goal of this series is twofold: first, to provide a set of readers on thematic topics for introductory or survey courses in bioethics or for courses with a particular theme or time limitation. When used in conjunction with a core text that provides the appropriate level of analytical framework, the readers in this series provide specific analysis of a set of issues that meets the professor's individual needs and interests. Second, each of the readers in this series is designed with the student in mind and aims to present seminal articles and case studies that help students focus more thoroughly and effectively on specific topics that flesh out the ethical issues at the core of bioethics.

Volumes in the Readings in Bioethics Series:

Reproductive Technologies
Genetic Technologies
Death and Dying
Health Care Policy
Genetics

Genetics: Science, Ethics, and Public Policy

A Reader

Edited by
Thomas A. Shannon

A SHEED & WARD BOOK
ROWMAN & LITTLEFIELD PUBLISHERS, INC.
Lanham • Boulder • New York • Toronto • Oxford

A SHEED & WARD BOOK

ROWMAN & LITTLEFIELD PUBLISHERS, INC.

Published in the United States of America
by Rowman & Littlefield Publishers, Inc.
A wholly owned subsidiary of The Rowman & Littlefield Publishing Group, Inc.
4501 Forbes Boulevard, Suite 200, Lanham, Maryland 20706
www.rowmanlittlefield.com

PO Box 317
Oxford
OX2 9RU, UK

British Library Cataloguing in Publication Information Available

Library of Congress Cataloging-in-Publication Data

Genetics : science, ethics, and public policy : a reader / edited by Thomas A.
Shannon.
    p. ; cm. — (Readings in bioethics)
"A Sheed & Ward book."
  Includes bibliographical references and index.
  ISBN 0-7425-3237-2 (cloth : alk. paper) — ISBN 0-7425-3238-0 (pbk. : alk.
paper)
  1. Genetics—Moral and ethical aspects. 2. Bioethics.
  [DNLM: 1. Genetics, Medical—ethics—Collected Works. 2. Bioethical
Issues—Collected Works. 3. Genetic Engineering—ethics—Collected Works. 4.
Genetic Screening—ethics—Collected Works. 5. Public Policy—Collected Works.
QZ 50 G32796 2004] I. Shannon, Thomas A. (Thomas Anthony), 1940- II. Series.
  QH438.7.G465 2004
  576.5—dc22
                            2004021101

Printed in the United States of America

♾️™ The paper used in this publication meets the minimum requirements of American
National Standard for Information Sciences—Permanence of Paper for Printed Library
Materials, ANSI/NISO Z39.48-1992.

To friends of many decades

MaryAnn and Jim Kenary
Jan and Maureen Woolhouse
Bernie and Mark Colborn
Sandy Shook
and in memory of
Ted Pat and Bob Sprunger

Contents

Acknowledgments   ix

Introduction   xi

1 Biotechnology and the Threat of a Posthuman Future   1
       Francis Fukuyama

2 Crossing Species Boundaries   11
       Jason Scott Robert and Françoise Baylis

3 Genetic Counseling and the Disabled: Feminism Examines
       the Stance of Those Who Stand at the Gate   33
       Annette Patterson and Martha Satz

4 The Natural Father: Genetic Paternity Testing,
       Marriage, and Fatherhood   59
       Gregory E. Kaebnick

5 Ethics of Preimplantation Diagnosis for a Woman Destined
       to Develop Early-Onset Alzheimer Disease   75
       Dena Towner and Roberta Springer Loewy

6 Procreation for Donation: The Moral and Political Permissibility
       of "Having a Child to Save a Child"   81
       Mark P. Aulisio, Thomas May, and Geoffrey D. Block

7   Population Screening in the Age of Genomic Medicine   97
        Muin J. Khoury, Linda L. McCabe, and
        Edward R. B. McCabe

8   Navigating Race in the Market for Human Gametes   115
        Hawley Fogg-Davis

9   How Can You Patent Genes?   131
        Rebecca S. Eisenberg

10 Monitoring Stem Cell Research   147
        The President's Council on Bioethics

11 Nuclear Transplantation, Embryonic Stem Cells,
        and the Potential for Cell Therapy   173
        Konrad Hochedlinger and Rudolf Jaenisch

Index   193

About the Editor and Contributors   199

Acknowledgments

Gratefully acknowledged are the publishers and authors of the works listed below for their permission to reprint their publications.

Francis Fukuyama, "Biotechnology and the Threat of a Posthuman Future." *The Chronicle of Higher Education* (22 March 2002): B7–B10. Permission granted by the author.

Jason Scott Robert and Françoise Baylis, "Crossing Species Boundaries." *American Journal of Bioethics* 3 (2000): 1–13. © 2003 by the Massachusetts Institute of Technology.

Annette Patterson and Martha Satz, "Genetic Counseling and the Disabled: Feminism Examines the Stance of Those Who Stand at the Gate," *Hypatia* 17(2002): 118–142. Permission granted by Indiana University Press.

Gregory E. Kaebnick, "The Natural Father: Genetic Paternity Testing, Marriage, and Fatherhood." *Cambridge Quarterly of Health Care Ethics* 13 (2004): 49–60. Reprinted with the permission of Cambridge University Press.

Dena Towner and Roberta Springer Loewy, "Ethics of Preimplantation Diagnosis for a Woman Destined to Develop Early-Onset Alzheimer Disease." *The Journal of the American Medical Association* 287 (2002): 1038–1040. Permission granted by the American Medical Association.

Mark P. Aulisio, Thomas May, and Geoffrey D. Block, "Procreation for Donation: The Moral and Political Permissibility of 'Having a Child to Save a Child.'" *Cambridge Quarterly of Healthcare Ethics* 10 (2001): 408–19. Reprinted with the permission of Cambridge University Press.

Muin J. Khoury, Linda L. McCabe, and Edward R. B. McCabe, "Population Screening in the Age of Genomic Medicine." *The New England Journal of Medicine* 348 (2003): 50–58. Copyright © 2003 Massachusetts Medical Society.

Hawley Fogg-Davis, "Navigating Race in the Market for Human Gametes." *The Hastings Center Report* 31 (2001): 13–21. Permission granted by the author and The Hastings Center.

Rebecca S. Eisenberg, "How Can You Patent Genes?" *American Journal of Bioethics*
2 (2002): 3–11. Adapted from *Who Owns Life?* ed. David Magnus, Arthur Caplan,
and Glenn McGee. Amherst, N.Y.: Prometheus Books, 2002. By permission of
Prometheus Books.

The President's Council on Bioethics, "Monitoring Stem Cell Research." Chapter 2
of *Current Federal Law and Policy.* Washington, D.C.: GPO.

Konrad Hochedlinger and Rudolf Jaenisch, "Nuclear Transplantation, Embryonic
Stem Cells, and the Potential for Cell Therapy." *The New England Journal of Medicine* 349 (2003): 275–86. Copyright © 2003 Massachusetts Medical Society.

Introduction

Since its introduction over a decade ago, the field of bioethics has grown exponentially. Not only has it become established as an academic discipline with journals and professional societies, it is covered regularly in the media and affects people everyday around the globe.

One important development in the field has been the informal division into clinical and institutional bioethics. Institutional bioethics has to do with the ethical dilemmas associated with the various institutions, the majority of which are providers of health care services. Delivery of health care, allocations of health care payments, mergers, and the closing or restricting services of certain departments or even of hospitals or clinics themselves are systemic questions involving a broad range of ethical agenda. On the clinical side, the bevy of usual suspects of ethical dilemmas is increasing in complexity as technology moves forward, new interventions are proposed, and fantasies become realities. Few, for example, thought that human cloning would become a serious clinical, public policy, and institutional debate in 2002.

One of the major consequences of this quantitative and qualitative debate is that providing resources for introductory or even specialized courses is becoming much more difficult. This is particularly difficult in the case of editing an anthology to complement a text that provides an analysis of the core ethical issues. There is simply too much material to put into a single anthology that is reasonable in price and convenient in size.

This series is an attempt to resolve the problem of a cumbersome and expensive anthology by providing a set of manageable and accessible readers on specific topics. Thus each reader in the series will be on a specific topic— reproductive technologies, genetic technologies, death and dying, health care policy—and will be about two hundred pages in length. This is to provide professors with flexibility in designing their courses. Ideally, professors will use a core text to analyze the primary ethical issues in bioethics and will use the readers in this series to examine specific problems and cases, thus providing flexibility in designing syllabi as well as providing variety in presenting the course.

The goal of this series is twofold: first, to provide a set of readers on thematic topics for introductory or survey courses in bioethics or for courses with a particular theme or time limitation. In addition to a core text that provides the appropriate level of analytical framework, the readers in this series provide specific analysis of a set of issues that meets the professor's needs and interests. Second, each of the readers in this series is designed with the student in mind and aims to present seminal articles and case studies that help students focus more thoroughly and effectively on specific topics that flesh out the ethical issues at the core of bioethics.

1

Biotechnology and the Threat of a Posthuman Future

Francis Fukuyama

I was born in 1952, right in the middle of the American baby boom. For any person growing up as I did in the middle decades of the 20th century, the future and its terrifying possibilities were defined by two books, George Orwell's *1984* (1949) and Aldous Huxley's *Brave New World* (1932).

The two books were far more prescient than anyone realized at the time, because they were centered on two different technologies that would in fact emerge and define the world over the next two generations. The novel *1984* was about what we now call information technology: Central to the success of the vast, totalitarian empire that had been set up over Oceania was a device called the telescreen, a wall-sized flat-panel display that could simultaneously send and receive images from each individual household to a hovering Big Brother.

Brave New World, by contrast, was about the other big technological revolution about to take place, that of biotechnology. Bokanovskification, the hatching of people not in wombs but, as we now say, in vitro; the drug soma, which gave people instant happiness; the feelies, in which sensation was simulated by implanted electrodes; and the modification of behavior through constant subliminal repetition and, when that didn't work, through the administration of various artificial hormones.

With at least a half-century separating us from the publication of these books, we can see that, while the technological predictions they made were startlingly accurate, the political predictions of the first book, *1984,* were entirely wrong. The year 1984 came and went, with the United States still locked in a cold-war struggle with the Soviet Union. That year saw the introduction of a new model of the IBM personal computer and the beginning of

1

what became the PC revolution. But instead of becoming an instrument of centralization and tyranny, it led to just the opposite: the democratization of access to information and the decentralization of politics. Instead of Big Brother watching everyone, people could use the PC and Internet to watch Big Brother, as governments everywhere were driven to publish more information on their own activities.

Just five years after 1984, in a series of dramatic events that would earlier have seemed like political science fiction, the Soviet Union and its empire collapsed, and the totalitarian threat that Orwell had so vividly evoked vanished. People were again quick to point out that those two events—the collapse of totalitarian empires and the emergence of the personal computer, as well as of other forms of inexpensive information technology, from TVs and radios to faxes and e-mail—were not unrelated.

The political prescience of the other great dystopia, *Brave New World,* remains to be seen. Many of the technologies that Huxley envisioned, like in vitro fertilization, surrogate motherhood, psychotropic drugs, and genetic engineering for the manufacture of children, are already here or just over the horizon. But this revolution has only just begun.

Of the nightmares evoked by the two books, *Brave New World*'s has always struck me as more subtle and more challenging. It is easy to see what's wrong with the world of *1984:* The protagonist, Winston Smith, is known to hate rats above all things, so Big Brother devises a cage in which rats can bite at Smith's face in order to get him to betray his lover. That is the world of classical tyranny, technologically empowered but not so different from what we have tragically seen and known in human history.

In *Brave New World,* by contrast, the evil is not so obvious because no one is hurt; indeed, this is a world in which everyone gets what he or she wants. In this world, disease and social conflict have been abolished; there is no depression, madness, loneliness, or emotional distress; sex is good and readily available. No one takes religion seriously any longer, no one is introspective or has unrequited longings, the biological family has been abolished, and no one reads Shakespeare. But no one (save John the Savage, the book's protagonist) misses those things, either, since everyone is happy and healthy.

Since the novel's publication, there have probably been several-million high-school essays written in answer to the question, "What's wrong with this picture?" The answer given (on papers that get A's, at any rate) usually runs something like this: The people in *Brave New World* may be healthy and happy, but they have ceased to be human beings. They no longer struggle, aspire, love, feel pain, make difficult moral choices, have families, or do any of the things that we traditionally associate with being human.

But while that kind of answer is usually adequate to satisfy the typical high-school English teacher, it does not probe nearly deeply enough. For one can go on to ask, What is so important about being a human being in the traditional way that Huxley defines it? After all, the human race today is the product of an evolutionary process that has been going on for millions of years, one that, with any luck, will continue well into the future. There are no fixed human characteristics, except for a general capability to choose what we want to be, to modify ourselves in accordance with our desires. So who is to tell us that being human and having dignity means sticking with a set of emotional responses that are the accidental byproduct of our evolutionary history?

Huxley is telling us, in effect, that we should continue to feel pain, be depressed or lonely, or suffer from debilitating disease, all because that is what human beings have done for most of their existence as a species. Certainly, no one ever got elected to Congress on such a platform. Instead of taking those characteristics and saying that they are the basis for "human dignity," why don't we simply accept our destiny as creatures who modify themselves?

Because nature itself, and in particular human nature, has a special role in defining for us what is right and wrong, just and unjust, important and unimportant. So our final judgment on "what's wrong" with Huxley's brave new world stands or falls with our view of how important human nature is as a source of values.

Huxley was right. The most significant threat posed by contemporary biotechnology is the possibility that it will alter human nature and thereby move us into a "posthuman" stage of history. That is important, because human nature exists, is a meaningful concept, and has provided a stable continuity to our experience as a species. It is, conjointly with religion, what defines our most basic values. Human nature shapes and constrains the possible kinds of political regimes. So a technology powerful enough to reshape what we are will have possibly malign consequences for liberal democracy and the nature of politics itself.

It may be that, as in the case of *1984,* we will eventually find biotechnology's consequences are completely and surprisingly benign. It may be that technology will in the end prove much less powerful than it seems today, or that people will be moderate and careful in their application of it. But one of the reasons I am not quite so sanguine is that biotechnology, in contrast to many other scientific advances, mixes obvious benefits with subtle harms in one seamless package.

Nuclear weapons and nuclear energy were perceived as dangerous from the start and, therefore, were subject to strict regulation from the moment the Manhattan Project created the first atomic bomb in 1945. There may be products of biotechnology that will be similarly obvious in the dangers they pose to mankind—for example, superbugs, new viruses, or genetically modified foods that produce toxic reactions. The more typical threats raised by biotechnology, however, are those captured so well by Huxley, and they are summed up in the title of an article by the novelist Tom Wolfe: "Sorry, but Your Soul Just Died."

Medical technology offers us, in many cases, a devil's bargain: longer life, but with reduced mental capacity; freedom from depression, together with freedom from creativity or spirit; therapies that blur the line between what we achieve on our own and what we achieve because of the levels of various chemicals in our brains. Consider the following three scenarios, all of which are distinct possibilities that may unfold over the next generation or two.

The first has to do with new drugs. As a result of advances in neuropharmacology, psychologists discover that human personality is much more plastic than formerly believed. In the future, knowledge of genomics permits pharmaceutical companies to tailor drugs very specifically to the genetic profiles of individual patients and to greatly minimize unintended side effects. Stolid people can become vivacious; introspective ones extroverted; you can adopt one personality on Wednesday and another for the weekend.

In the second scenario, advances in stem-cell research allow scientists to regenerate virtually any tissue in the body, and life expectancies are pushed well above 100 years. The only problem is that there are many subtle, and some not-so-subtle, aspects of human aging that the biotech industry hasn't quite figured out how to fix: People grow mentally rigid and increasingly fixed in their views as they age and, try as they might, they can't make themselves sexually attractive to each other and continue to long for partners of reproductive age. Worst of all, they refuse to get out of the way, not just of their children, but of their grandchildren and great-grandchildren.

In a third scenario, the wealthy routinely screen embryos before implantation to optimize the kind of children they have. You can increasingly tell the social background of a young person by his or her looks and intelligence. Human genes have been transferred to animals and even to plants for research purposes and to produce new medical products; and animal genes have been added to certain embryos to increase their physical endurance or resistance to disease. Scientists have not dared to produce a full-scale chimera, half human and half ape, though they could; but young people begin to suspect that classmates who do much less well than they do are in fact genetically not fully human. Because, in fact, they aren't.

Sorry, but your soul just died.

Toward the very end of his life, Thomas Jefferson wrote, "The general spread of the light of science has already laid open to every view the palpable truth, that the mass of mankind has not been born with saddles on their backs, nor a favored few booted and spurred, ready to ride them legitimately, by the grace of God." The political equality enshrined in the Declaration of Independence rests on the empirical fact of natural human equality.

We vary greatly as individuals and by culture, but we share a common humanity that allows every human being to potentially communicate with, and enter into a moral relationship with, every other human being on the planet. The ultimate question raised by biotechnology is, What will happen to political rights once we are able to, in effect, breed some people with saddles on their backs, and others with boots and spurs?

What should we do in response to biotechnology that, in the future, will mix great potential benefits with threats that are either physical and obvious or spiritual and subtle? The answer is obvious: We should use the power of the state to regulate it. And if that proves to be beyond the power of any individual nation-state, biotechnology needs to be regulated on an international basis. We need to start thinking concretely—now—about how to build institutions that can discriminate between good and bad uses of biotechnology and that can effectively enforce those rules both nationally and internationally.

That obvious answer is not obvious to many of the participants in the current biotechnology debate. The discussion remains mired, at a relatively abstract level, in the ethics of procedures like cloning or stem-cell research—divided into one camp that would like to permit everything, and another camp that would like to ban wide areas of research and practice. The broader debate is, of course, an important one, but events are moving so rapidly that we will soon need more practical guidance on how we can direct future developments so that the technology remains man's servant rather than his master. Since it seems very unlikely that we will either permit everything or ban research that is highly promising, we need to find a middle ground.

The creation of new regulatory institutions is not something that should be undertaken lightly, given the inefficiencies that surround all efforts at regulation. For the past three decades, there has been a commendable worldwide movement to deregulate large sectors of every nation's economy, from airlines to telecommunications and, more broadly, to reduce the size and scope of government. The global economy that has emerged as a result is a far more efficient generator of wealth and technological innovation. Excessive regulation in the past, however, led many to become instinctively hostile to state intervention in any form, and it is that knee-jerk aversion to regulation that will be one of the chief obstacles to getting human biotechnology under political control.

But it is important to discriminate: What works for one sector of the economy will not work for another. Information technology, for example, produces many social benefits and relatively few harms and, therefore, has

appropriately gotten by with a fairly minimal degree of government regulation. Nuclear materials and toxic waste, on the other hand, are subject to
strict national and international controls because unregulated trade in them
would clearly be dangerous.

One of the biggest problems in making the case for regulating human
biotechnology is the common view that, even if it were desirable to stop technological advance, it is impossible to do so. If the United States or any other
single country tries to ban human cloning or germ-line genetic engineering or
any other procedure, people who want to do those things will simply move to
a more favorable jurisdiction where such activities are permitted. Globaliza-
tion and international competition in biomedical research will ensure that
countries that hobble themselves by putting ethical constraints on their scientific communities or biotechnology industries will be punished.

The idea that it is impossible to stop or control the advance of technology
is simply wrong. We, in fact, control all sorts of technologies and many types
of scientific research: People are no more free to experiment in the development of new biological-warfare agents than they are to experiment on human
subjects without the latter's informed consent. The fact that there are some individuals or organizations that violate those rules, or that there are countries
where the rules are either nonexistent or poorly enforced, is no excuse for not
making the rules in the first place. People get away with robbery and murder,
after all, which is not a reason to legalize theft and homicide.

Putting in place a regulatory system that will permit societies to control human biotechnology will not be easy: It will require legislators in countries
around the world to step up to the plate and make difficult decisions on complex
scientific issues. The shape and form of the institutions designed to put into effect new rules are still wide open; designing them to be minimally obstructive of
positive developments while giving them effective enforcement capabilities is a
significant challenge. Even more challenging will be the creation of common
rules at an international level, the forging of a consensus among countries with
different cultures and views on the underlying ethical questions. But political
tasks of comparable complexity have been successfully undertaken in the past.

Many of the current debates over biotechnology are polarized between the
scientific community and those with religious commitments. I believe that
such polarization is unfortunate because it leads many to believe that the only
reason one might object to certain advances in biotechnology is out of religious belief. Particularly in the United States, biotechnology has been drawn
into the debate over abortion; many researchers feel that valuable progress is
being checked out of deference to a small number of anti-abortion fanatics.

But it is important to be wary of certain innovations in biotechnology for
reasons that have nothing to do with religion. The case might be called Aris-

totelian, not because I am appealing to Aristotle's authority as a philosopher, but because I take his mode of rational philosophical argument about politics and nature as a model. Aristotle argued, in effect, that human notions of right and wrong—what we today call human rights—were ultimately based on human nature. That is, without understanding how natural desires, purposes, traits, and behaviors fit together into a human whole, we cannot understand human ends or make judgments about right and wrong, good and bad, just and unjust.

Aristotle, together with his immediate predecessors Socrates and Plato, initiated a dialogue about the nature of human nature that continued in the Western philosophical tradition right up to the early modern period, when liberal democracy was born.

While there were significant disputes over what human nature was, no one contested its importance as a basis for rights and justice. Among the believers in natural rights were the American Founding Fathers, who based their revolution against the British crown on it. Nonetheless, the concept has been out of favor for the past century or two among academic philosophers and intellectuals.

Whatever academic philosophers and social scientists may think of the concept of human nature, the fact that there has been a stable human nature throughout human history has had very great political consequences. As Aristotle and every serious theorist of human nature has understood, human beings are by nature cultural animals, which means that they can learn from experience and pass on that learning to their descendants through nongenetic means. Hence human nature is not narrowly determinative of human behavior but leads to a huge variance in the way people raise children, govern themselves, provide resources, and the like. Mankind's constant efforts at cultural self-modification are what lead to human history, and to the progressive growth in the complexity and sophistication of human institutions over time.

The fact of progress and cultural evolution led many modern thinkers to believe that human beings were almost infinitely plastic—that is, that they could be shaped by their social environment to behave in open-ended ways. It is here that the contemporary prejudice against the concept of human nature starts. Many of those who believed in the social construction of human behavior had strong ulterior motives: They hoped to use social engineering to create societies that were just or fair according to some abstract ideological principle. Beginning with the French Revolution, the world has been convulsed with a series of Utopian political movements that sought to create an earthly heaven by radically rearranging the most basic institutions of society, from the family to private property to the state. Those movements crested in the 20th century, with the socialist revolutions that took place in Russia, China, Cuba, Cambodia, and elsewhere.

By the end of the century, virtually every one of those experiments had failed; in their place came efforts to create or restore equally modern, but less politically radical, liberal democracies. One important reason for that world-wide convergence on liberal democracy had to do with the tenacity of human nature. For while human behavior is plastic and variable, it is not infinitely so; at a certain point, deeply rooted natural instincts and patterns of behavior reassert themselves to undermine the social engineer's best-laid plans.

Many socialist regimes abolished private property, weakened the family, and demanded that people be altruistic to mankind in general rather than to a narrower circle of friends and family. But evolution did not shape human beings in that fashion. Individuals in socialist societies resisted the new institutions at every turn and, when socialism collapsed after the fall of the Berlin Wall in 1989, older, more familiar patterns of behavior reasserted themselves everywhere.

Political institutions cannot abolish either nature or nurture altogether and succeed. The history of the 20th century was defined by two opposite horrors: the Nazi regime, which said biology was everything, and Communism, which maintained that it counted for next to nothing. Liberal democracy has emerged as the only viable and legitimate political system for modern societies because it avoided either extreme, shaping politics according to historically created norms of justice while not interfering excessively with natural patterns of behavior.

There were many other factors affecting the trajectory of history, which I discussed in my book *The End of History and the Last Man.* One of the basic drivers of the human historical process has been the development of science and technology, which is what determines the horizon of economic-production possibilities and, therefore, a great deal of a society's structural characteristics. The development of technology in the late 20th century was particularly conducive to liberal democracy. That is not because technology promotes political freedom and equality per se—it does not—but because late-20th-century technologies (particularly those related to information) are what the political scientist Ithiel de Sola Pool has labeled technologies of freedom.

There is no guarantee, however, that technology will always produce such positive political results. As the more perceptive critics of the concept of the "end of history" have pointed out, there can be no end of history without an end of modern natural science and technology.

Not only are we not at an end of science and technology; we appear to be poised at the cusp of one of the most momentous periods of technological advance in history. Biotechnology and a greater scientific understanding of the human brain promise to have extremely significant political ramifications. As we discover not just correlations but actual molecular pathways between genes and traits like intelligence, aggression, sexual identity, criminality, alcoholism, and the like, it will inevitably occur to people that they can make use of the knowledge for particular social ends. That will play itself out as a series of ethical questions facing individual parents, and also as political issues that may someday come to dominate politics. If wealthy parents suddenly have open to them the opportunity to increase the intelligence of their children as well as that of all their subsequent descendants, then we have the makings not just of a moral dilemma but of a full-scale class war.

If we are worried about some of the long-term consequences of biotechnology, we can do something about it by establishing a regulatory framework to separate legitimate and illegitimate uses. The advance of technology is so rapid that we need to move quickly. What is important to recognize is that the challenge is not merely an ethical one but a political one as well. For it will be the political decisions that we make in the next few years concerning our relationship to this technology that determine whether we enter into a posthuman future and the potential moral chasm that such a future opens before us.

2

Crossing Species Boundaries

Jason Scott Robert and Françoise Baylis

INTRODUCTION

Crossing species boundaries in weird and wondrous ways has long interested the scientific community but has only recently captured the popular imagination beyond the realm of science fiction. Consider, for instance, the print and pictorial publicity surrounding the growth of a human ear on the back of a mouse;[1] the plight of Alba, artist Eduardo Kac's green-fluorescent-protein bunny stranded in Paris;[2] the birth announcement in *Nature* of ANDi, the first transgenic primate;[3] and, most recently, the growth of pigs' teeth in rat intestines[4] and miniature human kidneys in mice.[5]

But, bizarrely, these innovations that focus on discrete functions and organs are almost passé. As part of the project of harnessing the therapeutic potential of human stem cell research, researchers are now involved in creating novel interspecies whole organisms that are unique cellular and genetic admixtures (DeWitt 2002). A human-to-animal embryonic chimera is a being produced through the addition of human cellular material (such as pluripotent or restricted stem cells) to a nonhuman blastocyst or embryo. To give but four examples of relevant works in progress, Snyder and colleagues at Harvard have transplanted human neural stem cells into the forebrain of a developing bonnet monkey in order to assess stem cell function in development (Ourednik et al. 2001); human embryonic stem cells have been inserted into young chick embryos by Benvenisty and colleagues at the Hebrew University of Jerusalem (Goldstein et al. 2002); and most recently it has been reported that human genetic material has been transferred into rabbit eggs by Sheng (Dennis 2002), while Weissman and colleagues at Stanford University and StemCells, Inc., have created a mouse with a significant proportion of human stem cells in its brain (Krieger 2002).

Human-to-animal embryonic chimeras are only one sort of novel creature currently being produced or contemplated. Others include: *human-to-animal fetal or adult chimeras* created by grafting human cellular material to late-stage nonhuman fetuses or to postnatal nonhuman creatures; *human-to-human embryonic, fetal, or adult chimeras* created by inserting or grafting exogenous human cellular material to human embryos, fetuses, or adults (e.g., the human recipient of a human organ transplant, or human stem cell therapy); *animal-to-human embryonic, fetal, or adult chimeras* created by inserting or grafting nonhuman cellular material to human embryos, fetuses, or adults (e.g., the recipient of a xenotransplant); *animal-to-animal embryonic, fetal, or adult chimeras* generated from nonhuman cellular material whether within or between species (excepting human beings); *nuclear-cytoplasmic hybrids,* the offspring of two animals of different species, created by inserting a nucleus into an enucleated ovum (these might be intraspecies, such as sheep-sheep; or interspecies, such as sheep-goat; and, if interspecies, might be created with human or nonhuman material); *interspecies hybrids* created by fertilizing an ovum from an animal of one species with a sperm from an animal of another (e.g., a mule, the offspring of a he-ass and a mare); and *transgenic organisms* created by otherwise combining genetic material across species boundaries.

For this paper, in which we elucidate and explore the concept of species identity and the ethics of crossing species boundaries, we focus narrowly on the creation of interspecies chimeras involving human cellular material—the most recent of the transgressive interspecies creations. Our primary focus is on human-to-animal *embryonic* chimeras, about which there is scant ethical literature, though the scientific literature is burgeoning.

Is there anything ethically wrong with research that involves the creation of human-to-animal embryonic chimeras? A number of scientists answer this question with a resounding "no." They argue, plausibly, that human stem cell proliferation, (trans)differentiation, and tumorigenicity must be studied in early embryonic environments. For obvious ethical reasons, such research cannot be carried out in human embryos. Thus, assuming the research must be done, it must be done in nonhuman embryos—thereby creating human-to-animal embryonic chimeras. Other scientists are less sanguine about the merits of such research. Along with numerous commentators, they are quite sensitive to the ethical conundrum posed by the creation of certain novel beings from human cellular material, and their reaction to such research tends to be ethically and emotionally charged. But what grounds this response to the creation of certain kinds of part-human beings? In this paper we make a first pass at answering this question. We critically examine what we take to be the underlying worries about crossing species boundaries by referring to the creation of certain kinds of novel beings involving human cellular or genetic material. In turn, we highlight the limitations of each of these arguments. We then briefly hint at an alternative objection to the creation of certain novel beings that presumes a strong desire to avoid introducing moral confusion as regards the moral status of the novel being. In particular we explore the strong interest in avoiding any practice that would lead us to doubt the claim that humanness is a necessary (if not sufficient) condition for full moral standing.

SPECIES IDENTITY

Despite significant scientific unease with the notion of *species identity*, commonplace among biologists and commentators are the assumption that species have particular identities and the belief that the boundaries between species are fixed rather than fluid, established by nature rather than by social negotiation. Witness the ease with which biologists claim that a genome sequence of some organism—yeast, work, human—represents the identity of that species, its blueprint or, alternatively, instruction set. As we argue below, such claims mask deep conceptual difficulties regarding the relationship between these putatively representative species-specific genomes and the individual members of a species.

The ideas that natural barriers exist between divergent species and that scientists might someday be able to cross such boundaries experimentally fuelled debates in the 1960s and 1970s about the use of recombinant DNA technology (e.g., Krimsky 1982). There were those who anticipated the possibility of research involving the crossing of species boundaries and who considered this a laudable scientific goal. They tried to show that fixed species identities and fixed boundaries between species are illusory. In contrast those most critical of crossing species boundaries argued that there were fixed natural boundaries between species that should not be breached.

At present the prevailing view appears to be that species identity is fixed and that species boundaries are inappropriate objects of human transgression. The idea of fixed species identities and boundaries is an odd one, though, inasmuch as the creation of plant-to-plant[6] and animal-to-animal hybrids, either artificially or in nature, does not foster such a vehement response as the prospective creation of interspecies combinations involving human beings—no one sees rhododendrons or mules (or for that matter goat-sheep, or geep) as particularly monstrous (Dixon 1984). This suggests that the only species whose identify is generally deemed genuinely "fixed' is the human species. But, what is a *species* such that protecting its identity should be perceived by some to be a scientific, political, or moral imperative? This and similar questions about the nature of species and of species identities are important to address in the context of genetics and genomics research (Ereshefsky 1992; Claridge, Dawah, and Wilson 1997: Wilson 1999b).

Human beings (and perhaps other creatures) intuitively recognize species in the world, and cross-cultural comparative research suggests that people around the globe tend to carve up the natural world in significantly similar ways (Atran 1999). There is, however, no one authoritative definition of species. Biologists typically make do with a plurality of species concepts, invoking one or the other depending on the particular explanatory or investigative context.

One stock conception, propounded by Dobzhansky (1950) and Mayr (1940), among others, is the *biological species concept* according to which species are defined in terms of reproductive isolation, or lack of genetic exchange. On this view, if two populations of creatures do not successfully interbreed, then they belong to different species. But the apparent elegance and simplicity of this definition masks some important constraints: for instance, it applies only to those species that reproduce sexually (a tiny fraction of all species); moreover, its exclusive emphasis on interbreeding generates counterintuitive results, such as the suggestion that morphologically indistinguishable individuals who happen to live in neighboring regions but also happen never to interbreed should be deemed members of different species. (Imagine viewing populations of human beings "reproductively isolated" by religious intolerance as members of different species, and the biological species concept fails to pick out *Homo sapiens* as a discrete species comprising all human beings.)

Such results can be avoided by invoking other definitions of species, such as the *evolutionary species concept* advanced by G. G. Simpson and E. O. Wiley, which emphasizes continuity of populations over geological time: "a species is a single lineage of ancestral descendant populations of organisms which maintains its identity from other such lineages and which has its own evolutionary tendencies and historical fate" (Wiley 1978, 18; see also Simpson 1961). Unlike the biological species concept, this definition of species applies to both sexually and asexually reproducing creatures and also underscores shared ancestry and historical fate—and not merely capacity to interbreed—as what unifies a group of creatures as a species. The evolutionary species concept is by no means un-problematic, however, mainly because it is considerably more vague than the biological species concept, and so also considerably more difficult to operationalize.

A third approach to defining species has lately received considerable attention among philosophers of biology. This approach is known as the *homeostatic property cluster* view of species, advocated in different ways by Boyd

(1999), Griffiths (1999), and Wilson (1999a). Following Wilson (1999a, 197–99) in particular, the homeostatic property cluster view of species is properly understood as a thesis about natural kinds, of which a species is an instance. The basic idea is that a species is characterized by a cluster of properties (traits, say) no one of which, and no specific set of which, must be exhibited by any individual member of that species, but some set of which must be possessed by all individual members of that species. To say that these property clusters are "homeostatic" is to say that their clustering together is a systematic function of some causal mechanism or process; that an individual possesses any one of the properties in the property cluster significantly increases the probability that this individual will also possess other properties in the cluster. So the list of distinguishing traits is a property cluster, wherein the properties cluster as a function of the causal structure of the biological world. Of course, an outstanding problem remains, namely that of establishing the list of traits that differentiate species one from the other. Presumably this would be achieved by focusing on reproductive, morphological, genealogical, genetic, behavioral, and ecological features, no one of which is necessarily a universal property of the species and no set of which constitutes a species essence. We return below to the homeostatic property cluster view of species when we consider how best to characterize *Homo sapiens.*

To these definitions of species many more can be added: at present, there are somewhere between nine and twenty-two definitions of species in the biological literature.[7] Of these, there is no one species concept that is universally compelling. Accordingly, rather than asking the generic question, "How is 'species' defined?" it might be useful to focus instead on the narrower question "How is a species defined?" In response to the latter question Williams (1992) proposes that a species be characterized by a description comprising a set of traits differentiating that species from all others; it is no small task, however, to devise a satisfactory species description for any particular group of beings. Take, for example, *Homo sapiens.* Significantly, not even a complete sequence of *the* human genome can tell us what particular set of traits of *Homo sapiens* distinguishes human beings from all other species.

When molecular biologists first talked about mapping and sequencing the human genome, their goal was to construct the sequence of nucleotides in all the genes in all the chromosomes in the normal human body, The sequence was meant to serve as a reference point to which individual genomes could be compared in efforts to locate deviant genes implicated in phenotypic variation. As well, the sequence was meant to facilitate the study of gene function in development (often in comparison with the consensus genomes of organisms belonging to other species) and to establish historical relationships among organisms.

Two draft sequences of a "standard" or "typical" human genome were published in 2001, one produced under the auspices of the publicly-funded Human Genome Project (HGP), the other by Celera Genomics. The HGP's official genome is a composite of genetic information from tens or hundreds of human individuals, while Celera Genomics' official genome is a composite of genetic information from five individuals (but principally Craig Venter, Celera's former president; Wade 2002). The sequences are nonetheless supposed to be 99.9% identical to individual human genomes, and that 0.1% variation, in concert with environmental variations, is supposed to explain the immense diversity among human beings (for a recent statement of this position, see Plomin et al. 2002). But, excepting identical twins, every human genome is different from every other. Further, while one's maternal DNA may differ by 0.1% from one's paternal DNA, and one's own DNA may differ from that of any other individual by 0.1%, it is not the case that there is a certain part of an individual's genome that is 99.9% identical with every other human's genome. Although human beings might share 99.9% commonality at the genetic level, there is nothing as yet identifiable as *absolutely* common to all human beings. According to current biology, there is no genetic lowest common denominator, no genetic essence, "no single, standard, 'normal' DNA sequence that we all share" (Lewontin 1992, 36). The only way to determine how common the standard sequences are is to compare them with the actual sequences of a large number of individuals in an effort to detect conserved portions and polymorphisms; no one, though, is proposing such an endeavor. Even so, there is no way in which a single genome—not even Craig Venter's—can *represent* the immense genetic variability characteristic of *Homo sapiens* (Tauber and Sarkar 1992; Lloyd 1994; Robert 1998).

Moreover, comparative genomic research has thus far been of no help in establishing the boundary of human species identity. Much of "our" DNA is shared with a huge variety of apparently distantly related creatures (e.g., yeast, worms, mice). Indeed, given the evidence that all living things share a common ancestor, there is little (if any) uniquely human DNA.[8] More strikingly perhaps, though human beings are morphologically and behaviorally vastly different from chimpanzees, we differ genomically from chimps by no more than 1.2–1.6% (Allen 1997; Marks 2002; Enard et al. 2002; Olson and Varki 2003). Further, the surprisingly small number of genes in the sequenced human genome, as compared to original estimates, offers a serious blow to the idea of human uniqueness at the genomic level (Claverie 2001). Finally, there is no comfort to be found in the assessment that a tiny number of physical, chemical, genetic, and developmental accidents made human history possible. In sum, even though biologists are able to identify a particular string of nucleotides as human (as distinct from, say, yeast or even chimpanzee), the unique identity of the human species cannot be established through genetic or genomic means.

WHAT *IS* HOMO SAPIENS?

What, then, is *Homo sapiens?* Though clearly there is no one authoritative definition of species, notions of "species essences" and "universal properties of species" persist, always in spirit if not always in nature, in discussions about breaching species boundaries. For this reason, on occasion, attempts to define *Homo sapiens* are reduced to attempts to define *human nature.* This is a problem, however, insofar as the literature exhibits a wide range of opinion on the nature of *human nature;* indeed, many of the competing conceptions of *human nature* are incommensurable (for a historical sampling of views, see Trigg 1988). On one view the claim that there is such a thing as human nature is meant to be interpreted as the claim that all members of *Homo sapiens* are essentially the same. But since everything about evolution points toward variability and no essential sameness, this would appear to be an inherently problematic claim about human nature (Hull 1986). One way of avoiding this result is to insist that talk of human nature is not about essential sameness but rather about universality and then to explain universality in terms of distinct biological attributes—a functional human nervous system, a human anatomical structure and physiological function, or a human genome (Campbell, Glass, and Chatland 1998). A classic example of the latter strategy, explaining universality genetically, appears in an article on human nature by Eisenberg (1972), who writes that "one trait common to man everywhere is language; in the sense that only the human species displays it, the capacity to acquire language must be genetic" (126).[9] In this brief passage Eisenberg moves from the claim that language is a human universal, to the claim that the ability to have a language is unique and species specific, to the claim that this capacity is genetic (Hull 1986). But, of course, language is not a human universal—some human beings neither speak nor write a language, and some are born with no capacity whatsoever for language acquisition. Yet, in a contemporary context, no one would argue that these people, simply by virtue of being nonverbal and/or illiterate, are not members of the same species as the rest of us.[10]

And therein lies the rub. We all know a human when we see one, but, really, that is all that is known about our identity as a species. Of course we all know that human beings are intelligent, sentient, emotionally-complex creatures. We all know the same of dolphins, though. And, of course, not all human beings are intelligent, sentient, or emotionally complex (for instance, those who are comatose); nevertheless, most among us would still consider them human.

The homeostatic property cluster approach to species avoids the problem of universality but at the possible expense of retaining an element of essentialism. Recall that, according to the homeostatic property cluster view, membership in a species is not determined by possession of *any particular* individual homeostatically clustered property (or *any particular sets* of them) but rather by possession of *some* set of homeostatically clustered properties. Nevertheless, although possession of property *x* (or of property set *x-y-z)* is not *necessary* for species membership, possession of *all* the identified homeostatically clustered properties is *sufficient* for membership, which suggests that a hint of essentialism persists (Wilson 1999a).

This is an ironic result, inasmuch as essentialism in biology is vanishingly rare. This is because essentialism—or at least stock conceptions of essentialism according to which a species is identified by essential intrinsic properties—is at odds with evolutionary biology.[11] Significantly, commentators of all stripes tend to revert to essentialist thinking when pondering the locus of humanity. This might be because of a persistent folk essentialism, reflecting "a way of thinking about living systems whose continuing grip on us is explained by the fact that it develops long before we are exposed to scientific biology" (Griffiths 2002, 77). It might also be because the very idea of a "locus of humanity" is always already an essentialist idea.

MORAL UNREST WITH CROSSING SPECIES BOUNDARIES[12]

As the above discussion of species identity makes clear, there is no consensus on what exactly is being breached with the creation of interspecies beings. As against what was once commonly presumed, there would appear to be no such thing as fixed species identities. This fact of biology, however, in no way undermines the reality that fixed species exist independently as moral constructs. That is, notwithstanding the claim that biologically species are fluid, people believe that species identities and boundaries are indeed fixed and in fact make everyday moral decisions on the basis of this belief. (There is here an analogy to the recent debate around the concept of race. It is argued that race is a biologically meaningless category, and yet this in no way undermines the reality that fixed races exist independently as social constructs and they continue to function, for good or, more likely, ill, as a moral category.) This gap between science and morality requires critical attention.

Scientifically, there might be no such thing as fixed species identities or boundaries. Morally, however, we rely on the notion of fixed species identities and boundaries in the way we live our lives and treat other creatures, whether in decisions about what we eat or what we patent. Interestingly, there is dramatically little appreciation of this tension in the literature, leading us to suspect that (secular) concern over breaching species boundaries is in fact concern about something else, something that has been mistakenly characterized in the essentialist terms surveyed above. But, in a sense, this is to be expected. While a major impact of the human genome project has been to show us quite clearly how similar we human beings are to each other and to other species, the fact remains that human beings are much more than DNA and moreover, as we have witnessed throughout the ages, membership within the human community depends on more than DNA. Consider, for example, the not-so-distant past in which individual human beings of a certain race, creed, gender, or sexual orientation were denied moral standing as members of the human community. By appealing to our common humanity, ethical analysis and social activism helped to identify and redress what are now widely seen as past wrongs.

Although in our recent history we have been able to broaden our understanding of what counts as human, it would appear that the possible permeability of species boundaries is not open to public debate insofar as novel part-human beings are concerned. Indeed, the standard public-policy response to any possible breach of human species boundaries is to reflexively introduce moratoriums and prohibitions.[13]

But why should this be so? Indeed, why should there be *any* ethical debate about the prospect of crossing species boundaries between human and nonhuman animals? After all, hybrids occur naturally, and there is a significant amount of gene flow between species in nature.[14] Moreover, there is as yet no adequate biological (or moral) account of the distinctiveness of the species *Homo sapiens* serving to capture all and only those creatures of human beings born. As we have seen, neither essentialism (essential sameness, genetic or otherwise) nor universality can function as appropriate guides in establishing the unique identity of *Homo sapiens*. Consequently, no extant species concept justifies the erection of the fixed boundaries between human beings and nonhumans that are required to make breaching those boundaries morally problematic.[15] Despite this, belief in a fixed, unique, human species identity persists, as do moral objections to any attempt to cross the human boundary—whatever that might be.

According to some, crossing species boundaries is about human beings playing God and in so doing challenging the very existence of God as infallible, all-powerful, and all-knowing. There are, for instance, those who believe that God is perfect and so too are all His creations. This view, coupled with the religious doctrine that the world is complete, suggests that our world is perfect. In turn, perfection requires that our world already contains all possible creatures. The creation of new creatures—hybrids or chimeras—would confirm that there are possible creatures that are not currently found in the world, in which case "the world cannot be perfect; therefore God, who made the world, cannot be perfect; but God, by definition is perfect; therefore God could not exist" (Morriss 1997, 279).[16] This view of the world, as perfect and complete, grounds one sort of opposition to the creation of human-to-animal chimeras.

As it happens, however, many do not believe in such a God and so do not believe it is wrong to "play God." Indeed, some would argue further that not only is it *not* wrong to play God, but rather this is exactly what God enjoins us to do. Proponents of this view maintain that God "left the world in a state of imperfection so that we become His partners"—his co-creators (Breitowitz 2002, 327).

Others maintain that combining human genes or cells with those of nonhuman animals is not so much about challenging God's existence, knowledge, or power, as it is about recognizing this activity as inherently unnatural, perverse, and so offensive. Here the underlying philosophy is one of repugnance. To quote Kass (1998),

> repugnance revolts against the excesses of human willfulness, warning us not to transgress what is unspeakably profound. Indeed in this age in which . . . our given human nature no longer commands respect . . . repugnance may be the only voice left that speaks up to defend the central core of humanity. (19)

For many, the mainstay of the argument against transgressing species "boundaries" is a widely felt reaction of "instinctive hostility" (Harris 1998, 177) commonly known as the "yuck factor." But in important respects repugnance is an inchoate emotive objection to the creation of novel beings that requires considerable defense. If claims about repugnance are to have any moral force, the intuitions captured by the "yuck" response must be clarified. In the debate about the ethics of creating novel beings that are part human, it is not enough to register one's intuitions. Rather, we need to be able to clearly identify and critically examine these intuitions, recognizing all the while that they derive "from antecedent commitment to categories that are themselves subject to dispute" (Stout 2001, 158).

A plausible "thin" explanation for the intuitive "yuck" response is that the creation of interspecies creatures from human materials evokes the idea of bestiality—an act widely regarded as a moral abomination because of its degrading character. Sexual intimacy between human and nonhuman animals typically is prohibited in law and custom, and some, no doubt, reason from the prohibition on the erotic mixing of human and nonhuman animals to a prohibition on the biotechnological mixing of human and nonhuman cellular or genetic material. There are important differences, however. In the first instance the revulsion is directed toward the shepherd who lusts after his flock and acts in a way that makes him seem (or actually be) less human (Stout 2001, 152). In the second instance the revulsion is with the purposeful creation of a being that is neither uncontroversially human nor uncontroversially nonhuman.

A more robust explanation for the instinctive and intense revulsion at the creation of human-to-animal beings (and perhaps some animal-to-human beings) can be drawn from Douglas's work on taboos (1966). Douglas suggests that taboos stem from conceptual boundaries. Human beings attach considerable symbolic importance to classificatory systems and actively shun anomalous practices that threaten cherished conceptual boundaries. This explains the existence of well-entrenched taboos, in a number of domains, against mixing things from distinct categories or having objects/actions fall outside any established classification system. Classic examples include the Western response to bi-sexuality (you can't be both heterosexual and homosexual) and intersexuality. Intersexuality falls outside the "legitimate" (and exclusive) categories of male and female, and for this reason intersex persons have been carved to fit into the existing categories (Dreger 2000). Human-to-animal chimeras, for instance, are neither clearly animal nor clearly human. They obscure the classification system (and concomitant social structure) in such a way as to constitute an unacceptable threat to valuable and valued conceptual, social, and moral boundaries that set human beings apart from all other creatures. Following Stout, who follows Douglas, we might thus consider human-to-animal chimeras to be an abomination. They are anomalous in that they "combine characteristics uniquely identified with separate kinds of things, or at least fail to fall unambiguously into any recognized class." Moreover, the anomaly is loaded with social significance in that interspecies hybrids and chimeras made with human materials "straddle the line between *us* and *them*" (Stout 2001, 148). As such, these beings threaten our social identity, our unambiguous status as human beings.

But what makes for unambiguous humanness? Where is the sharp line that makes for the transgression, the abomination? According to Stout, the line must be both sharp and socially significant if trespassing across it is to generate a sense of abomination: "An abomination, then, is anomalous or ambiguous with respect to some system of concepts. And the repugnance it causes depends on such factors as the presence, sharpness, and social significance of conceptual distinctions" (Stout 2001, 148). As we have seen, though, there is no biological sharp line: we have no biological account of unambiguous humanness, whether in terms of necessary and sufficient conditions or of homeostatic property clusters. Thus it would appear that in this instance abomination is a social and moral construct.

Transformative technologies, such as those involved in creating interspecies beings from human material, threaten to break down the social dividing line between human beings and nonhumans. Any offspring generated through the pairing of two human beings is by natural necessity—reproductive, genetic, and developmental necessity—a human. But biology now offers the prospect of generating offspring through less usual means; for instance, by transferring nuclear DNA from one cell into an enucleated egg. Where the nuclear DNA and the enucleated egg (with its mitochondrial DNA) derive from organisms of different species, the potential emerges to create an interspecies nuclear-cytoplasmic hybrid.

In 1998 the American firm Advanced Cell Technology (ACT) disclosed that it had created a hybrid embryo by fusing human nuclei with enucleated cow oocytes. The goal of the research was to create and isolate human embryonic stem cells. But if the technology actually works (and there is some doubt about this) there would be the potential to create animal-human hybrids (ACT 1998; Marshall 1998; Wade 1998). Any being created in this way would have DNA 99% identical with that of the adult from whom the human nucleus was taken; the remaining 1% of DNA (i.e., mitochondrial DNA) would come from the enucleated animal oocyte. Is the hybrid thus created simply part-human and part-nonhuman animal? Or is it unequivocally human or unequivocally animal (see Loike and Tendlet 2002)? These are neither spurious nor trivial questions. Consider, for example, the relatively recent practice in the United States of classifying octoroons (persons with one-eighth Negro blood; the offspring of a quadroon and a white person) as black. By analogy, perhaps 1% animal DNA (i.e., mitochondrial DNA) makes for an animal.[17]

A more complicated creature to classify would be a human-to-animal chimera created by adding human stem cells to a nonhuman animal embryo. It has recently been suggested that human stem cells should be injected into mice embryos (blastocysts) to test their pluripotency (Dewitt 2002). If the cells were to survive and were indeed pluripotent, they could contribute to the formation of every tissue. Any animal born following this research would be a chimera—a being with a mixture of (at least) two kinds of cells. Or, according to others, it would be just a mouse with a few human cells. But what if those cells are in the brain, or the gonads (Weissman 2002)? What if the chimeric mouse has human sperm? And what if that mouse were to mate with a chimeric mouse with human eggs?

All of this is to say that when faced with the prospect of not knowing whether a creature before us is human and therefore entitled to all of the rights typically conferred on human beings, we are, as a people, baffled.

One could argue further that we are not only baffled but indeed fearful. Hybrids and chimeras made from human beings represent a metaphysical threat to our self-image. This fear can be explained in both historical and contemporary concerns. Until the end of the eighteenth century the dominant Western worldview rested on the idea of the Great Chain of Being. The world was believed to be an ordered and hierarchical place with God at the top, followed by angels, human beings, and various classes of animals on down through to plants and other lesser living matter (Lovejoy 1970; see also Morriss 1997). On this worldview human beings occupied a privileged place between the angels and all nonhuman animals. In more recent times, though the idea of the Great Chain of Being has crumbled, the reigning worldview is still that human beings are superior to animals by virtue of the human capacity for reason and language. Hybrids and chimeras made from human materials blur the fragile boundary between human beings and "unreasoning animals," particularly when one considers the possibility of creating "reasoning" nonhuman animals (Krieger 2002). But is protecting one's privileged place in the world solid grounds on which to claim that hybrid—or chimera—making is intrinsically or even instrumentally unethical?

MORAL CONFUSION

Taking into consideration the conceptual morass of species-talk, the lack of consensus about the existence of God and His role in Creation, healthy skepticism about the "yuck" response, and confusion and fear about obscuring, blurring, or breaching boundaries, the question remains as to why there should be any ethical debate over crossing species boundaries. We offer the following musings as the beginnings of a plausible answer, the moral weight of which is yet to be assessed.

All things considered, the engineering of creatures that are part human and part nonhuman animal is objectionable because the existence of such beings would introduce inexorable moral confusion in our existing relationships with nonhuman animals and in our future relationships with part-human hybrids and chimeras. The moral status of nonhuman animals, unlike that of human beings, invariably depends in part on features other than species membership, such as the intention with which the animal came into being. With human beings the intention with which one is created is irrelevant to one's moral status. In principle it does not matter whether one is created as an heir, a future companion to an aging parent, a sibling for an only child, or a possible tissue donor for a family member. In the case of human beings, moral status is categorical insofar as humanness is generally considered a necessary condition for moral standing. In the case of nonhuman animals, though, moral status is contingent on the will of regnant human beings. There are different moral obligations, dependent on social convention, that govern our behavior toward individual nonhuman animals depending upon whether they are bred or captured for food (e.g., cattle), for labor (e.g., oxen for subsistence farming), for research (e.g., lab animals), for sport (e.g., hunting), for companionship (e.g., pets), for investment (e.g., breeding and racing), for education (e.g., zoo animals), or whether they are simply cohabitants of this planet. In addition, further moral distinctions are sometimes drawn between "higher" and "lower" animals, cute and ugly animals, useful animals and pests, all of which add to the complexity of human relationships with nonhuman animals.

These two frameworks for attributing moral status are clearly incommensurable. One framework relies almost exclusively on species membership in *Homo sapiens* as such, while the other relies primarily on the will and intention of powerful "others" who claim and exercise the right to confer moral status on themselves and other creatures, For example, though some (including ourselves) will argue that the biological term *human* should not be conflated with the moral term *person,* others will insist that all human beings have an inviolable moral right to life simply by virtue of being human. In sharp contrast, a nonhuman animal's "right to life" depends entirely upon the will of some or many human beings, and this determination typically will be informed by myriad considerations.

It follows that hybrids and chimeras made from human materials are threatening insofar as there is no clear way of understanding (or even imagining) our moral obligations to these beings—which is hardly surprising given that we are still debating our moral obligations to some among us who are undeniably biologically human, as well as our moral obligations to a range of nonhuman animals. If we breach the clear (but fragile) *moral* demarcation line between human and nonhuman animals, the ramifications are considerable, not only in terms of sorting out our obligations to these new beings but also in terms of having to revisit some of our current patterns of behavior toward certain human and nonhuman animals.[18] As others have observed (e.g., Thomas 1983), the separateness of humanity is precarious and easily lost; hence the need for tightly guarded boundaries.

Indeed, asking—let alone answering—a question about the moral status of part-human interspecies hybrids and chimeras threatens the social fabric in untold ways; countless social institutions, structures, and practices depend upon the moral distinction drawn between human and nonhuman animals. Therefore, to protect the privileged place of human animals in the hierarchy of being, it is of value to embrace (folk) essentialism about species identities and thus effectively trump scientific quibbles over species and over the species status of novel beings. The notion that species identity can be a fluid construct is rejected, and instead a belief in fixed species boundaries that ought not to be transgressed is advocated.

An obvious objection to this hypothesis is that, at least in the West, there is already considerable confusion and lack of consensus about the moral status of human embryos and fetuses, patients in a persistent vegetative state, sociopaths, nonhuman primates, intelligent computers, and cyborgs. Given the already considerable confusion that exists concerning the moral status of this range of beings, there is little at risk in adding to the confusion by creating novel beings across species boundaries. Arguably, the current situation is already so morally confused that an argument about the need to "avoid muddying the waters further" hardly holds sway.[19]

From another tack, others might object that confusion about the moral status of beings is not new. There was a time when many whom we in the West now recognize as undeniably human—for example, women and blacks—were not accorded this moral status. We were able to resolve this moral "confusion" (ongoing social discrimination not withstanding) and can be trusted to do the same with the novel beings we create.

Both of these points are accurate but in important respects irrelevant. Our point is not that the creation of interspecies hybrids and chimeras adds a huge increment of moral confusion, not that there has never been confusion about the moral status of particular kinds of beings, but rather that the creation of novel beings that are part human and part nonhuman animal is sufficiently threatening to the social order that for many this is sufficient reason to prohibit any crossing of species boundaries involving human beings. To do otherwise is to have to confront the possibility that humanness is neither necessary nor sufficient for personhood (the term typically used to denote a being with full moral standing, for which many—if not most—believe that humanness is at least a necessary condition).

In the debate about the ethics of crossing species boundaries the pivotal question is: Do we shore up or challenge our current social and moral categories? Moreover, do we entertain or preclude the possibility that humanness is not a necessary condition for being granted full moral rights? How we resolve these questions will be important not only in determining the moral status and social identity of those beings with whom we currently coexist (about whom there is still confusion and debate), but also for those beings we are on the cusp of creating. Given the social significance of the transgression we contemplate embracing, it behooves us to do this conceptual work now, not when the issue is even more complex—that is, once novel part-human beings walk among us.

CONCLUSION

To this point we have not argued that the creation of interspecies hybrids or chimeras from human materials should be forbidden or embraced. We have taken no stance at all on this particular issue. Rather, we have sketched the complexity and indeterminacy of the moral and scientific terrain, and we have highlighted the fact that despite scientists' and philosophers' inability to precisely define species, and thereby to demarcate species identities and boundaries, the putative fixity of putative species boundaries remains firmly lodged in popular consciousness and informs the view that there is an obligation to protect and preserve the integrity of human beings and *the* human genome. We have also shown that the arguments against crossing species boundaries and creating novel part-human beings (including interspecies hybrids or chimeras from human materials), though many and varied, are largely unsatisfactory. Our own hypothesis is that the issue at the heart of the matter is the threat of inexorable moral confusion.

With all this said and done, in closing we offer the following more general critique of the debate about transgressing species boundaries in creating part-human beings. The argument, insofar as there is one, turns something like this: species identities are fixed, not fluid; but just in case, prohibiting the transgression of species boundaries is a scientific, political, and moral imperative. The scientific imperative is prudential, in recognition of the inability to anticipate the possibly dire consequences for the species *Homo sapiens* of building these novel beings. The political imperative is also prudential, but here the concern is to preserve and protect valued social institutions that presume pragmatically clear boundaries between human and nonhuman animals. The moral imperative stems from a prior obligation to better delineate moral commitments to both human beings and animals before undertaking the creation of new creatures for whom there is no apparent a priori moral status.

As we have attempted to show, this argument against transgressing species boundaries is flawed. The first premise is not categorically true—there is every reason to doubt the view that species identity is fixed. Further, the scientific, political, and moral objections sketched above require substantial elaboration. In our view the most plausible objection to the creation of novel interspecies creatures rests on the notion of moral confusion—about which considerably more remains to be said.

ACKNOWLEDGMENTS

An early version of this paper was presented at the 2001 meeting of the American Society for Bioethics and Humanities, Nashville, Tenn. A significantly revised version of this paper was presented at the 2002 meeting of the International Association of Bioethics, Brasilia, Brazil. We are grateful to both audiences for engaging our ideas. Additional thanks are owed to Brad Abernethy, Vaughn Black, Fern Brunger, Josephine Johnston, Jane Maienschein, Robert Perlman, and Janet Rossant for helpful comments on interim drafts. Research for this paper was funded in part by grants from the Canadian Institutes of Health Research (CIHR) made independently to JSR and FB, a grant from the CIHR Institute of Genetics to JSR, and a grant from the Stem Cell Network (a member of the Networks of Centers of Excellence program) to FB.

NOTES

1. See, e.g., Mooney and Mikso (1999); and the *Scientific American Frontiers* coverage of "The Bionic Body," available from: http://www.pbs.org/saf/1107/features/body.htm. See also Bianco and Robey (2001).

2. See the bibliography of media coverage at http:/www.ekac.org/transartbiblio .html.

3. A sample headline from *The Independent* (London): "How a Glowing Monkey Will Help Cure Disease" (12 January 2001; available from: http://www.independent .co.uk/story.jsp?story=49841). See also Chan et al. (2001); and Harris (2001).

4. "Scientists Grow Pig Teeth in Rat Intestines" is available from: http:// www.laurushealth.com/HealthNews/reuters/NewsStory0926200224.htm. See also Young et al. (2002).

5. "Human Kidneys Grown in Mice" is available from: http://news.bbc.co.uk/ 2/hi/health/2595397.stm. See also Dekel et al. (2003).

6. A possible exception is the creation of genetically modified crops. But here the arguments are based on human health and safety concerns, as well as on political opposition to monopolistic business practices, rather than on concern for the essential identity of plant species.

7. Kitcher (1984) and Hull (1999) each discuss nine concepts. Mayden (1997) discusses twenty-two.

8. In fact through studies in comparative genomics biologists have demonstrated horizontal transfer or genes between lineages, suggesting a remarkable fluidity of species "boundaries" at the genomic level. Some of this literature is reviewed in Doolittle (1999).

9. Other examples are everywhere to be found in commentaries on the human genome project.

10. And even were language a human universal *par excellence,* there is simply no basis for the assumption that invariability (universality) and genetics must be connected. See Hull (1986); and Oyama (2000).

11. This case is usually made in terms of Mayr's distinction between (non-Darwinian) typological thinking and (Darwinian) population thinking (Mayr 1959). For a useful

account of Mayr's distinction, see Sober (1980). Griffiths (1999) attempts to resurrect an alternative account of essentialism compatible with Darwinism, wherein he deals not with intrinsic essential properties but rather extrinsic (relational) ones. We will not discuss this effort here, nor will we address the view that typological thinking has an important role to play in contemporary evolutionary biology in approaching the evolution of form (Love 2003).

12. Given our suggestion that the notion of species boundaries is problematic, at least biologically speaking, it might seem odd for us to continue using the language of "crossing species boundaries." We offer two defenses: first, the language is commonly used, especially to capture some sort of moral demarcation line (see below); second, we intend the notion, biologically, in a limited sense. Consider any individual human. That individual human contains a genome, a specifically human genome; call this genome H. Next, consider some nonhuman animal, or even a plant; call this genome not-H. Next, consider the application of standard genetic manipulation techniques to isolate a particular functional stretch of DNA from this specific not-H genome. Finally, consider the application of standard gene transfer techniques to insert (across "species boundaries," as we here understand the term) the gene from not-H into H, via the germ line. Some of the offspring of the bearer of genome H would thereafter contain genomes in which the gene from not-H appears. The bearer of H and her/his offspring would thus be interspecies beings (in the limited biological sense intended).

13. See, for example, §6(2)(b) Infertility (Medical Procedures) Act 1984 (Victoria, Australia); §3(2)(a)–(b) and §3(3)(b) Human Fertilisation and Embryology Act 1990 (United Kingdom); and Article 25 Bill containing rules relating to the use of gametes and embryos (Embryo Bill), September 2000 (the Netherlands). See also Annas, Andrews, and Isasi (2002).

14. A particularly well-documented example of gene flow between species is Darwin's finches in the Galapagos Islands. For a recent account, see Grant and Grant (2002).

15. A possible objection is that the biological species concept could in fact do the required work: human beings do not successfully interbreed with mice or moose, and so the boundary is established. We do not find the objection compelling. Whether human beings can in fact successfully interbreed with mice or moose is an open empirical question; while it does not happen in nature, it might happen artificially in the ways noted at the outset of this paper. The artificiality of such reproduction does not render it of a different kind, though. Human beings requiring reproductive technologies in order to breed are nonetheless human, they nonetheless reproduce, and they nonetheless generate offspring who are unquestionably human. So, the biological species concept cannot be used to discount the potential artificial creation of hybrids or of chimeras as a matter of breaching fixed species boundaries.

16. Note that Morriss does not subscribe to such a position.

17. Mitochondrial DNA is not insignificant DNA. Like nuclear DNA it codes for functions.

18. Animal-rights advocates might object to the creation of part-human hybrids on the grounds that this constitutes inappropriate treatment of animals solely to further

human interests. Obviously, proponents of such a perspective will not typically have a prior commitment to the uniqueness and "dignity" of human beings. For this reason we do not pursue this narrative here.

19. This objection was raised for us by Vaughan Black.

REFERENCES

Advanced Cell Technology. 1998. Advanced Cell Technology announces use of nuclear replacement technology for successful generation of human embryonic stem cells. Press release, 12 November. Available from: http://www.advancedcell.com/pr_11-12-1998.html.

Allen, B. 1997. The chimpanzee's tool. *Common Knowledge* 6(2): 34–54.

Annas, G. J., L. B. Andrews, and R. M. Isasi. 2002. Protecting the endangered human: Toward an international treaty prohibiting cloning and inheritable alterations. *American Journal of Law & Medicine* 28:151–78.

Atran, S. 1999. The universal primacy of generic species in folk biological taxonomy: Implications for human biological, cultural, and scientific evolution. In *Species: New interdisciplinary essays,* ed. R. A. Wilson, 231–61. Cambridge: MIT Press.

Bianco, P., and P. G. Robey. 2001. Stem cells in tissue engineering. *Nature* 414:118–21.

Boyd, R. 1999. Homeostasis, species, and higher taxa. In *Species: New interdisciplinary essays,* ed. R. A. Wilson, 141–85. Cambridge: MIT Press.

Breitowitz, V. 2002. What's so bad about human cloning? *Kennedy Institute of Ethics Journal* 12:325–41.

Campbell, A., K. G. Glass, and L. C. Charland. 1998. Describing our "humanness": Can genetic science alter what it means to be "human"? *Science and Engineering Ethics* 4:413–26.

Chan, A. W. S., K. V. Chong, C. Martinovich, C. Simerly, and G. Schatten. 2001. Transgenic monkeys produced by retroviral gene transfer into mature oocytes. *Science* 291:309–12.

Claridge, M. F., H. A. Dawah, and M. R. Wilson, eds. 1997. *Species: The units of biodiversity.* London: Chapman and Hall.

Claverie, J. M. 2001. What if there are only 30,000 human genes? *Science* 291:1255–57.

Dekel, B., T. Burakova, F. D. Arditti et al. 2003. Human and porcine early kidney precursors as a new source for transplantation. *Nature Medicine* 9:53–60.

Dennis, C. 2002. China: Stem cells rise in the East. *Nature* 419:334–36.

DeWitt, N. 2002. Biologists divided over proposal to create human-mouse embryos. *Nature* 420:255.

Dixon, B. 1984. Engineering chimeras for Noah's ark. *Hastings Center Report* 10:10–12.

Dobzhansky, T. 1950. Mendelian populations and their evolution, *American Naturalist* 84:401–18.

Doolittle, W. F. 1999. Lateral genomics. *Trends in Genetics* 15(12): M5–M8.

Douglas, M. 1966. *Purity and danger.* London: Routledge and Kegan Paul.

Dreger, A. D. 2000. *Hermaphrodites and the medical invention of sex.* Cambridge: Harvard University Press.

Eisenberg, L. 1972. The *human* nature of human nature. *Science* 176:123–28.

Enard, W., P. Khaitovich, J. Klose et al. 2002. Intra- and interspecific variation in primate gene expression patterns. *Science* 296:340–43.

Ereshefsky, M., ed. 1992. *The units of evolution: Essays on the nature* of *species.* Cambridge: MIT Press.

Goldstein, R. S., M. Drukker, B. E. Reubinoff, and N. Benvenisty. 2002. Integration and differentiation of human embryonic stem cells transplanted to the chick embryo. *Developmental Dynamics* 225:80–86.

Grant, P. R., and B. R. Grant. 2002. Unpredictable evolution in a 30-year study of Darwin's finches. *Science* 296:633–35.

Griffiths, P. 1999. Squaring the circle: Natural kinds with historical essences. In *Species: New interdisciplinary essays,* ed. R. A. Wilson, 209–28. Cambridge: MIT Press.

———. 2002. What is innateness? *The Monist* 85:70–85.

Harris, J. 1998. *Clones, genes, and immortality: Ethics and the genetic revolution.* New York: Oxford University Press.

Harris, R. 2001. Little green primates. *Current Biology* 11:R78–R79.

Hull, D. L. 1986. On human nature. In *Proceedings of the Biennial Meeting of the Philosophy of Science Association* 2:3–13.

———. 1999. On the plurality of species: Questioning the party line. In *Species: New interdisciplinary essays,* ed. R. A. Wilson, 23–48. Cambridge: MIT Press.

Kass, L. J. 1998. The wisdom of repugnance. In *The ethics of human cloning,* by L. J. Kass and J. A. Wilson, 3–59. Washington: AEI Press.

Kitcher, P. 1984. Species. *Philosophy of Science* 51:308–33.

Krieger, L. M. 2002. Scientists put a bit of man into a mouse. *Mercury News,* 8 December. Available from: http://www.bayarea.com/mid/mercurynews/4698610.htm.

Krimsky, S. 1982. *Genetic alchemy: The social history of the recombinant DNA controversy.* Cambridge: MIT Press.

Lewontin, R. C. 1992. The dream of the human genome. *New York Review of Books,* 28 May, pp. 31–40.

Lloyd, E. A. 1994. Normality and variation: The Human Genome Project and the ideal human type. In *Are genes us? The social consequences of the new genetics,* ed. C. F. Cranor, 99–112. New Brunswick, NJ: Rutgers University Press.

Loike, J. D., and M. D. Tendler. 2002. Revisiting the definition of *Homo sapiens. Kennedy Institute of Ethics Journal* 12:343–50.

Love, A. C. 2003. Evolutionary morphology, innovation, and the synthesis of evolutionary and developmental biology. *Biology & Philosophy* 18:309–45.

Lovejoy, A. O. 1970. *The great chain of being: A Study of the history of an idea.* Cambridge: Harvard University Press.

Marks, J. 2002. *What it means to be 98% chimpanzee: Apes, human beings, and their genes.* Berkeley: University of California Press.

Marshall, E. 1998. Claim of human-cow embryo greeted with skepticism. *Science* 282:1390–91.

Mayden, R. L. 1997. A hierarchy of species concepts: The denouement in the saga of the species problem. In *Species: The units of biodiversity,* ed. M. F. Claridge, H. A. Dawah, and M. R. Wilson, 381–424. London: Chapman and Hall.

Mayr, E. 1940. Speciation phenomena in birds. *American Naturalist* 74:249–78.

———. 1959. Typological versus populational thinking. In *Evolution and the Diversity of Life,* by E. Mayr, 26–29. Cambridge: Harvard University Press, 1976.

Mooney, D. J., and A. G. Mikos. 1999. Growing new organs. *Scientific American* 280:38–43.

Morriss, P. 1997. Blurred boundaries. *Inquiry* 40:259–90.

Olson, M. V., and A. Varki. 2003. Sequencing the chimpanzee genome: Insights into human evolution and disease. *Nature Reviews Genetics* 4:20–28.

Ourednik, V., J. Ourednik, J. D. Flax et al. 2001. Segregation of human neural stem cells in the developing primate forebrain. *Science* 293:1820–24.

Oyama, S. 2000. *The ontogeny of information: Developmental systems and evolution,* rev. ed. Durham: Duke University Press.

Plomin, R., J. C. Defries, I. W. Craig, P. McGuffin, and J. Kagan, eds. 2002. *Behavioral genetics in the post-genomic era.* Washington: American Psychological Association.

Robert, J. S. 1998. Illich, education, and the Human Genome Project: Reflections on paradoxical counter-productivity. *Bulletin of Science, Technology, and Society* 18:228–39.

Simpson, G. G. 1961. *Principles of animal taxonomy.* New York: Columbia University Press.

Sober, E. 1980. Evolution, population thinking, and essentialism. *Philosophy of Science* 47:350–83.

Stout, J. 2001. *Ethics after Babel: The languages of morals and their discontents.* Boston: Beacon Books, 1988. Reprint, in expanded form and with a new postscript, Princeton: Princeton University Press.

Tauber, A. I., and S. Sarkar. 1992. The Human Genome Project: Has blind reductionism gone too far? *Perspectives in Biology and Medicine* 35:220–35.

Thomas, K. 1983. *Man and the natural world: Changing attitudes in England, 1500–1800.* London: Allen Lane.

Trigg, R. 1988. *Ideas of human nature: An historical introduction.* Oxford, U.K.: Basil Blackwell.

Wade, N. 1998. Researchers claim embryonic cell mix of human and cow. *New York Times,* 12 November, p. A1. Available from: http://query.nytimes.com/search/article-page.html?res=9C04E3D71731F931A25752C1A96E958260.

———. 2002. Scientist reveals genome secret: It's his. *New York Times,* 27 April. Available from: http://www.nytimes.com/2002/04/27/science/27GENO.html.

Weissman, I. 2002. Stem cells: Scientific, medical, and political issues. *New England Journal of Medicine* 346:1576–79.

Wiley, E. O. 1978. The evolutionary species concept reconsidered. *Systematic Zoology* 27:17–26.

Williams, M. B. 1992. Species: Current usages. In *Keywords in evolutionary biology,* ed. E. F. Keller and E. A. Lloyd, 318–23. Cambridge: Harvard University Press.

Wilson, R. A. 1999a. Realism, essence, and kind: Resuscitating species essentialism? In *Species: New interdisciplinary essays,* ed. R. A. Wilson, 187–207. Cambridge: MIT Press.

———. 1999b. *Species: New interdisciplinary essays,* Cambridge: MIT Press.

Young, C. S., S. Terada, J. P. Vacanti et al. 2002. Tissue engineering of complex tooth structures on biodegradable polymer scaffolds. *Journal of Dental Research* 81:695–700.

3

Genetic Counseling and the Disabled: Feminism Examines the Stance of Those Who Stand at the Gate

Annette Patterson and Martha Satz

Imperfection is the essence of being organic and alive. Organic life is vulnerable; it inevitably ends in disintegration. This is part of its beauty.

—Joan Tollifson, "Imperfection Is a Beautiful Thing"

The present essay examines the crucial role of genetic counseling in determining society's attitude toward those with disabling conditions, the consequent ethical responsibilities and imperatives arising from such a position, and the contributions feminism may offer to ease the ethical burdens and untangle the dilemmas of the genetic counselor. For, as disability rights advocates argue and this paper affirms, the genetic counselor's attitudes and actions influence the material situation as well as the self-concept of those with disabilities. Certainly this discussion has urgency. With the completion of the Human Genome Project and the burgeoning fascination with DNA in the popular imagination, the scientific and cultural atmosphere is poised to launch the genetic counselor on a trajectory of ever-increasing significance. Indeed, over the last two decades, the prenatal landscape has changed drastically. In years past, genetic testing meant checking for the presence of a few chromosomal conditions such as Down syndrome. Geneticists understood relatively little about the human genome and had limited ability to predict the presence of, or predisposition to, genetic conditions. However, with the completion of the Human Genome Project, science has an ever-increasing ability to predict the future.

Geneticists may soon be able to hand prospective parents a list of genetic conditions/predispositions present in a given fetus at a very early stage in pregnancy. Parents will then be asked (and encouraged) to consider what level of disease/disability is acceptable and manageable for them and their families. Though genetic tests are requested and performed by doctors and lab technicians, in most cases, it is the genetic counselor that presents genetic information and presides over the discussion that proceeds from such testing. This, then, places genetic counselors in a critical position because they undoubtedly have, whether acknowledged or not, a strong impact on the perception of genetic conditions and on prenatal decisions following from those perceptions.

The field of genetic counseling has from its inception been self-conscious about the ethical import of its position. Haunted by the specter and history of eugenics, genetic counseling has been governed by the cardinal principle and moral imperative of nondirectiveness. Yet, until relatively recently, this controlling ideal has remained largely unexamined. Current literature does reflect doubt among counselors and others about whether nondirective counseling is either theoretically or practically possible and expresses additional concern about the extent to which the "nondirective" counselor simply replicates the biases of the larger society.[1] Furthermore, disability activists argue that whether or not the counselor can achieve nondirective prenatal counseling, the process of prenatal counseling, by its very nature, presupposes an implicit bias to abort selectively any fetus deemed "defective."

Literature by disability advocates emphasizes the power wielded by genetic counselors in this process. Serving both as purveyors of genetic information and as guides in decision-making, genetic counselors often preside over prenatal sessions where parents are considering whether to continue or terminate a pregnancy. This evaluation process has profound implications for society in shaping attitudes about what constitutes a "life . . . worth living" (Morris 1991, 8) and, potentially, the provisions society will make for those with disabilities. Thus, counselors stand at a crucial nexus between the disability community and the community at large because their own views of disability, as well as their own moral values, directly or indirectly affect clinical decisions and by extension, societal values. Because they occupy such a critical position, genetic counselors have a unique ethical responsibility to explore their attitudes and beliefs concerning disability, both conscious and unconscious, and to recognize the potential power of their stance toward disability in shaping societal views and public policy. And, particularly if genetic counselors do not have first-hand experience of disability, they have the ethical mandate to employ the varied perspectives of disabled individuals in conveying a picture of what it means to be disabled.

It is worthwhile to note here from a feminist perspective that the genetic counseling transaction is largely one between and among women. Most genetic counselors are in fact female (an estimated 95 percent); mothers, of course, make the final legal decision concerning the termination of a pregnancy; and, in our society, mothers shoulder most of the care of a disabled child.

Disability rights advocates have a strong political position to exert regarding their interest in such transactions, and in general, regarding the status of those who are disabled. Their political movement, as other liberatory movements before them, asserts that an apparent "natural" hierarchy, in this case separating the abled from the disabled, is in fact one largely constructed by societal practice. They contend that the difficulties presented by a given disabling condition often have more to do with the way society defines and responds to these conditions rather than with the inherent "limitations" of the conditions themselves. They further argue that genetic counselors participate in this "construction" by attempting to define and explain the nature of a particular condition often without themselves having any significant experience of disability or interaction with disabled individuals. Because genetic counselors occupy an authoritative position and because their clients often have little experience with disability, the definition they present becomes a frame of reference for prenatal decision-making. With this in mind, how should genetic counselors respond to the troubling claim that the very enterprise of genetic counseling is inevitably political, and to the conclusion that may reasonably be derived from such a position, that each act of genetic counseling is a political act in which one woman directly influences the reproductive decision of another woman?

Clearly, feminist thought has much to contribute to the problems and dilemmas of genetic counseling. In recent years, feminist thought has become accustomed to reformulating itself in response to the claims of racial, ethnic, and class concerns exerted by various contingents of women. The voices of disability activists exert yet another important perspective to demonstrate the inequitable power relationships that manifest themselves as influences upon women's reproductive decisions. Furthermore, an ethic of care has been developed within the context of feminist thought, an ethic that emphasizes the specific nature of individual relationships and responsibilities. Insights from such ethicists as Carol Gilligan (1982) and Nel Noddings (1984) can shed light on the genetic counseling relationship and can provide tools to explore the potential relationships within the family of a disabled person.

Indeed, it is feminist thought that has most elaborately articulated the theory of standpoint epistemology, a view that is particularly helpful in elucidating the essential role of the disability rights perspective in genetic counseling. Standpoint epistemology contends that knowledge claims are always socially situated and that failure by dominant groups to interrogate beliefs arising from their social situation leaves them in an epistemologically disadvantaged position, that is, one that distorts. Some would argue that genetic counselors, for the most part, occupy such a position because many in the field, however well-intentioned, are woefully unaware of both the actual contours of disabled people's lives and their attitudes regarding their conditions. Nevertheless, operating from such a privileged and distorting situation, genetic counselors provide information to prospective parents about specific conditions, thereby structuring the parents' view of that condition and consequently influencing decisions regarding the future existence of a person with such a condition.

Standpoint epistemology and the methodological tools that proceed from it can be brought to bear on the various moral problems of genetic counseling. It can illuminate the issue of nondirectiveness as it occurs in the genetic counseling situation, the nature of optimal communication in the counseling process, the future education of genetic counselors, and the relation of those with disabilities to genetic counselors and to the wider community. Concurring with the viewpoint that knowledge is inevitably situated, especially in the particular context of disability, Erik Parens and Adrienne Asch (2000a) sum up the results of a Hastings Institute project on the topic of prenatal testing and disability rights that included participants with diverse backgrounds: "If any of us ever did, no one in our group can any longer imagine having a view from nowhere. Those of us with disabilities know that our particular experience of discrimination colors our critique of prenatal testing. Those of us who used prenatal testing before or during the project appreciate that this experience colors our responses to those critiques. Those of us who are parents sometimes found ourselves justifying our own parental attitudes, whereas those of us who are not parents sometimes asked ourselves whether becoming parents might change our views about what constitutes an admirable 'parental attitude'" (Parens and Asch 2000a, ix–x).

With regard to such situated knowledge, feminist philosophers have, in the last two decades, elaborated a theory that may well aid genetic counselors in integrating the perspectives of those with disabilities into the counseling process and also give them a justification for doing so. As Nancy Hartsock remarks, "There are some perspectives on society [here we may read, the able-bodied] from which, however well-intentioned one may be, the real relations of humans with each other and with the natural world are not visible" (1983, 159). The corollary to this view is that the lives at the "bottom" of social hierarchies, the experience and lives of marginalized people (in this case, the disabled), can provide particularly fruitful vantage points from which to identify problem areas of a specific enterprise in need of exploration, such as prenatal testing and associated genetic counseling.

In an article reflecting upon standpoint epistemology, the feminist philosopher Sandra Harding offers guidance for how the privileged may incorporate the marginalized in their actions and thinking, a model potentially useful for genetic counselors in their practice. Harding offers Patricia Hill Collins's (see 1990) discussion of African American feminist thought as an ideal applicable to other marginalized groups: ". . . this approach challenges members of dominant groups to make themselves 'fit' to engage in collaborative . . . enterprises with marginal peoples. Such a project requires learning to listen attentively to marginalized people; it requires educating oneself about their histories, achievements, preferred social relations, and hopes for the future. . . it requires critical examination of the dominant institutional beliefs and practices that systematically disadvantage them; it requires critical self-examination to discover how one unwittingly participates in generating disadvantage to them . . . and more" (1993, 68).

Such a feminist analysis has much to offer the genetic counselor because it suggests a method for integrating the perspectives of members of the disabled community. Through this process, the counselor will come to understand that her knowledge is inevitably situated, and so in her counseling she must actively work against her bias—in most cases, the bias of the privileged, able-bodied person. Genetic counseling may then replace the ideal of nondirectiveness with the ideal of a self-conscious critique of bias. But let us consider the role of nondirectiveness within the field of genetic counseling.

NONDIRECTIVENESS

At the core of the field of genetic counseling lies a paradox epitomized in the injunction that the counselor be nondirective. Ann Platt Walker perhaps unwittingly suggests part of the conflict by her choice of language in *A Guide to Genetic Counseling* (1998): "Adherence to a nonprescriptive (often referred to as 'nondirective') approach is perhaps the most defining feature of genetic counseling. The philosophy stems from a firm belief that genetic counseling should—insofar as is possible—be devoid of any eugenic motivation" (8).

The phrase "insofar as is possible" within the dashes reveals a problem, suggesting its author's concern that the enterprise of genetic counseling may by its nature be unable to free itself totally from a eugenic cast. In "Implications of the Human Right to Life" (1983) Leon Kass more forcefully interjects such a fear: ". . . persons afflicted with certain diseases will never be capable of living the full life of a human being. . . . There is no reason to keep them alive. This standard, I would suggest, is the one which most physicians and genetic counselors appeal to in their hearts, no matter what they say or do. . . . Why else would they have developed genetic counseling?" (400). The

underlying contention is that the act of offering information about the defects
and disabilities of a fetus in a medical setting, where abortion is offered as an
alternative, in itself makes abortion more of, to borrow William James's term,
a "live option" (see James 1897) than it may have been before the information was imparted. Further complicating this issue is the fact that pregnant
women often undergo such testing without becoming fully aware that they
may refuse to do so. Increasingly, obstetricians do not "offer" pregnant
women prenatal testing but rather suggest it as a routine part of prenatal care.
Thus, without any formal discussion or truly informed consent, many women,
without ever realizing they have done so, embark upon a course that may end
in a discussion of pregnancy termination due to suspected fetal abnormality.
In "Collective Silences, Collective Fictions" (1994) Nancy Press and Carole
Browner talk about the move from risk-based to population-based screening
that has "occurred with little explicit comment on the profound change it represents" (201). Press and Browner argue that this quiet revolution in prenatal
testing has been fueled by the acceptance, on the part of both physicians and
pregnant women, of a "collective fiction: the presentation of AFP screening
as a simple and routine part of prenatal care" (203). This "collective fiction"
allows all to remain silent about what the eventual consequences of the test
may be. It allows all parties ". . . to remain silent about issues on which there
is no societal consensus—such as the appropriateness of aborting fetal anom-
alies and the eugenic implications of the practice" (214). This shift in prenatal practice came easily and without much opposition, they argue, because it
serves everyone's interests: "It serves the state public health program's purpose of trying to reduce the incidence of neural tube defects, health care
providers' desire to limit their legal liability, and women's complex needs to
be reassured about the outcome of their pregnancy and to leave open but un-
contemplated the option of terminating an affected pregnancy" (203).

This shift in the justification for prenatal testing is subtle but disturbing,
and though it can scarcely be termed eugenic with regard to intent, it may be
so in effect, and serious moral objections may be raised to the process. What
is even more disturbing is the practice of some physicians who insist that patients carrying fetuses diagnosed with certain conditions attend genetic counseling sessions even after their patients have declared that they would not
seek an abortion under any circumstances and therefore did not wish to see a
counselor (see Saxton 1984 and Felker 1994). While physicians may be motivated by legal considerations and a professional responsibility to meet the
anticipated medical and psycho-social needs of the family, it is understand-
able that patients may interpret such "coercion" as implying that any rational
person, if fully informed, would have an abortion under these circumstances.
As one woman quoted by Marsha Saxton puts it: "As I was examined and in-

terviewed by several different disciplines, I was left with the impression that continuing a pregnancy [of a fetus with spinal bifida] such as mine was an unusual thing to do. It seemed as though every time I turned around another physician was asking me whether or not anyone had discussed my 'options' with me. 'Options' has clearly become a euphemism for abortion" (Saxton 2000, 157). Simply put, counseling itself may offer a message.

And this process will only intensify. Interestingly and ironically, as genetic information becomes more wide-ranging, underlying concerns about eugenics will become even more complex. Since parents will be privy to a wide array of genetic information comprising various conditions and predispositions, the fate of all fetuses that are the subject of a genetic counseling session will become the subject of an explicit decision, not simply fetuses diagnosed with certain conditions. Thus, the necessity for genetic counselors to contextualize the information provided becomes even more critical.

Let us suppose, for example, that a particular fetus has one gene that will predispose it to cleft palate, and another that predisposes it to obesity or depression. Undoubtedly, the genetic counselor will remind her clients that all fetuses will have some imperfections or predisposition to some unfavorable condition, that conditions such as cleft palate are correctable, and that prior information about such predisposing factors will help prevent the most negative consequences of such a gene. In other words, the genetic counselor will provide information that will mitigate the direst conclusions inferred from such information. However, in the case of what society perceives as a disabling condition, such as Down syndrome or Achondroplasia, will the counselor do the same? Or is the bias against these conditions so entrenched, albeit invisible, both in the greater society and in the genetic counseling community, that such ameliorating contextualization is unlikely to occur?

A SPECIAL CASE AND ITS IMPLICATIONS FOR NONDIRECTIVENESS

A dilemma recently arising within genetic counseling throws into bold relief the possible bias against such disabling conditions within the genetic counseling community. In the past few years, genetic counselors have reportedly encountered individuals who seek information about possible assistance in having children with a particular genetic disorder through the use of medical technology. Thus far, requests for this type of information have come from individuals with Achondroplasia wishing to have achondroplastic children or Deaf parents who want their children to be Deaf as well (Davis 1997b, 7). Parents in both groups argue that their genetic "conditions" are important and valuable aspects of their identities and that these conditions form both their cultures and their communities. They want their children to share in this identity and to take their place in the community of their parents.

The Deaf are particularly vocal about their views. They argue that deafness confers a cultural status, not a state of disability. From their perspective, being Deaf "opens the child up to membership in the Deaf community, which has a rich history, language and value system of its own" (Crouch 1997, 18). In an article entitled "Defiantly Deaf," published in the *New York Times* Magazine, 28 August 1994, Andrew Solomon explains that ". . . being Deaf is a culture and a source of pride. . . . A steadily increasing number of deaf people have said that they would not choose to be hearing. To them, the word 'cure'—indeed the whole notion of deafness as pathology—is an anathema." Similar arguments have been proffered by members of the achondroplastic community, who affirm their right to have children that reflect their physical likeness, their experiences, and their values. They assert that physical difference is not synonymous with abnormality and that the birth of a child with Achondroplasia should not be interpreted as a tragic event.

However, opponents charge that in these cases, satisfying the needs and desires of the parents means unfair "diminishment" of the child. In "Genetic Dilemmas and the Child's Right to an Open Future" (1997b), Dena Davis argues against allowing parents to choose hereditary deafness for their children: "A decision, made before a child is even born, that confines her forever to a narrow group of people and a limited choice of careers, so violates the child's right to an open future that no genetic counseling team should acquiesce in it" (14). Davis's statement is striking for a number of reasons. But before we consider it, let us make some preliminary clarifications. Davis and other opponents of this practice are not arguing against procreative freedom for the Deaf or for individuals with Achondroplasia who attempt to have children while accepting the attendant risks. This right is widely acknowledged. Rather, they object to prospective parents actively choosing for their unborn children what the medical establishment has defined as "disease." We would argue that Davis's position is highly significant. Even though it addresses an extreme case in genetic counseling, admittedly affecting only a small number of persons, the view's significance stems from the fact that it clearly advocates abandoning the nondirective principle, which negates parental autonomy in reproductive decision-making. Some reproductive decisions, Davis argues, are simply immoral, and genetic counselors should not aid and abet them. Such a case reveals that some in the field would openly admit that the guiding principle of nondirectiveness has its limits. Furthermore, Davis defends the view that this reproductive decision is immoral by affirming that, contrary to the claim of some disability activists, deafness is a disability rather than a socially constructed disadvantage.[2] She thus assumes that hearing is intrinsically preferable to deafness, thereby affirming her position, the position of a hearing person, held in opposition to that of the potential parents, who are members of the Deaf culture.

Davis's stance is grounded in the view that choosing deafness irrevocably restricts the child's right to an "open future," a right to which Davis, following Joel Feinberg's position, declares every child is entitled (Feinberg 1992, 84). In the case of the Deaf, Davis declares, "marriage partners, conversation partners, vocations, and avocations are severely limited." She adds, "Yes, one can think of cultural minorities about whom the same could be said—e.g. the Amish or very Orthodox Jews—but these children can change their minds as adults and a significant percentage do so" (Davis 1997a, 3).

We would maintain that Davis's argument concerning this unusual and rare genetic counseling case derives its significance from its very appeal to common sense. It is intellectually attractive, cogent, and representative of a mainstream and institutionalized stance against disability among medical professionals, a stance vulnerable to cultural criticism and to standpoint epistemology. Such critiques reveal Davis's bias, a bias that has the most sweeping and possibly devastating impact upon those with disabilities.

Davis's anchoring principle, Feinberg's notion of the "right to an open future," is certainly at first blush an alluring one. However, as a fundamental desideratum, this value is itself strongly culturally laden. Unrestricted freedom and maximal choices are, we may realize on second thought, quintessentially mainstream American values. Other cultures and even cultural communities within this country may emphasize other values, ties restrictive by their very nature, to family, community, culture, and even deity. Consider Asian cultures or, as Davis herself mentions, the Amish. But in the context of the Amish, Davis, in her discussion of a Supreme Court case, privileges the premise of an "open future" against the claims of the Amish community to shape their young people, which is, in our view, to reassert the premise of the dominant culture. Feinberg, in discussing this Supreme Court decision regarding the Amish, remarks: "An impartial decision would assume only that education should equip the child with the knowledge and skills that will help him choose whichever sort of life best fits his native endowment and matured disposition. It should send him out into the adult world with as many open opportunities as possible, thus maximizing his chance for self-fulfillment" (1992, 84). Such comments reveal the cultural assumptions of the viewpoint. In this time and in this place, the dominant culture links multiplicity of opportunity with self-fulfillment, the latter itself a culturally laden notion. William Ruddick (2000) further reveals the cultural bias of such a view. As he develops his position, he begins a statement "This principle steers between the Amish or Hasidic project of literally reproducing the parents' lives in their children. . . ." (101). With this off-hand clause,

Ruddick consigns all the Amish to one life and all Hassidim to a second. He assumes, in the case of the Amish, that all rural lives are the same, and in the case of the Hassidim, that living within a religious community negates the variability of their lives, in spite of the fact that members of this community pursue widely diverse professions.[3] With such an off-the-cuff remark, a culturally biased viewpoint reveals itself.

Furthermore, while the right to an open future seems a worthwhile goal, upon thought and examination, the notion itself collapses. "An open future" is an ideal and abstract notion, which breaks down as soon as the world impinges.[4] No child has an open future. Biology, even amongst the able-bodied, mitigates against it. The genetic lottery dictates that some children, no matter how hard they try, cannot become nuclear physicists, ballerinas, NFL quarterbacks, or philosophers. Also, as every thoughtful parent comes to realize, every child-rearing decision closes many doors. For example, bringing up a child within a particular religion prevents that child from having her most basic and emotional associations with the rituals and symbols of other religions. As an adult, she may convert to another religion, but primitive powerful emotional associations may make such a conversion problematic. We are generally aware of the critical window at which language may be acquired, but are less cognizant that other components of one's cognitive equipment are formed at crucial junctures as well. A mundane example may illustrate the problematic nature of the "open future" criterion.

When our friends' talented and precocious daughter reached the age of five, her swimming instructor approached her parents to tell them that their daughter had the potential to become an Olympic swimmer. He recommended that they institute a serious schedule of lessons and practice for their daughter. Her parents, Hungarian intellectual Jews, laughed to themselves, "Who wants their daughter to be an Olympic swimmer?" They had a generally negative attitude toward athletics, often remarking to their children that many animals could run faster than humans, so why try to compete? In their view, it was the quintessentially human that should be developed. Two years later, when this same daughter exhibited an equally remarkable talent for the violin, they hired the best teacher in the city, and the mother practiced with her daughter three hours a day. One might argue that becoming either an Olympic swimmer or a concert violinist requires serious training from early childhood, and that whatever choice parents make restricts that child's open future, her ability to attain either of these goals.

However, even assuming that the notion of an "open future" is a problematic, if not vacuous, one, the argument still may be offered that choosing deafness for one's child *unduly* restricts her, prevents her from a wide variety of choices. But such an argument begs rather than decides the question of whether deafness is a disability, whether membership in a culture numerically small when compared to the majority is disadvantageous rather than advantageous. Those in the majority culture may *assume* that it is more advantageous to live in the majority culture than in the minority culture, but without recognizing their cultural assumption as such, clothe that assumption in a philosophical argument such as entitlement to an open future. Davis would bolster her argument by asserting that asymmetry is involved, that a hearing person can join the Deaf culture, but the deaf cannot join hearing culture. But, again, that is not quite true. Although hearing persons can, for example, learn ASL, it is not clear that they can acquire the visual enhancement that those who speak it as a first language will have, thus losing out on many of the advantages of the culture. (For example, see Sacks 1989, 69-73.)

What emerges from Davis's argument, an argument that most in the genetic counseling community and the community at large would accept, is that deafness is more restrictive and thus inferior to hearing. It is an argument, we should notice, that is opposed by many of those who are deaf, who associate themselves with Deaf culture. It is at just such a juncture that standpoint epistemology would be instructive. By viewing such an argument, and the genetic counseling that would proceed from it, from the standpoint of those in the Deaf culture, we might be able to see that such a stance, rather than describing the world as it is, merely affirms a majoritarian cultural prejudice.

Clearly, Davis's view casts the experience of being deaf or achondroplastic in negative terms, assuming that no cultural benefit gained can override the sociological and medical risks. It also dismisses the notion that some aspects of life with Achondroplasia or hereditary deafness may actually enhance individual identity and life experience. Evidently, this is a view strongly influenced by the medical model of health and disease, which equates "normal" with "optimal" and sees any deviation from the norm as unfortunate (Davis and Bradley 1997, 69). Given the view of disability embedded within Davis's argument, one may wonder whether the ameliorating perspective of the disabled would be brought to bear on counseling sessions conducted by genetic counselors who share Davis's viewpoint—and whether, and to what extent, experiences are available to genetic counselors in general to dislodge them from this viewpoint.

A Personal Narrative

To illustrate this point, one author of the present paper offers a personal narrative, an event that was, in fact, the genesis of this paper.

At the 1999 genetic counseling convention in Oakland, California, a woman with Achondroplasia talked about her experience of having children.

She recounted in detail the story of her first pregnancy, clearly indicating that she and her husband (himself achondroplastic) had hoped for a child with Achondroplasia. Rather than perceiving their "condition" as negative, as something to be avoided, they believed that raising a child with Achondroplasia would be more comfortable and feasible than the alternative scenario. She talked about the birth of her son and described the challenges presented by his Achondroplasia. What followed was an account of repeated hospitalizations, multiple surgeries, and painful interventions. Now seven years old, he has never walked and has had three surgeries to place rods along his spinal column. All have failed. Wanting more children, but exhausted by their previous ordeal, they decided to adopt a child with Achondroplasia from Russia. This too was disappointing, as the child was plagued by severe emotional problems. Still wishing for a "normal" parenting experience, they decided to try one last time to have a baby biologically their own. Again, the speaker made it clear that they preferred a child with Achondroplasia. Their third child was born with Achondroplasia, but fortunately has had relatively few health problems. She expressed great satisfaction in having a "normal" child who did not require constant medical attention. She finished her talk and opened the floor for questions.

The convention hall fell silent. I was silent as well, but inwardly in turmoil. The story disturbed me. The fact that she wished for an affected child in the face of her previous nightmare seemed reprehensible to me. I could not understand why she would take such a position, knowing that affected status would increase the likelihood of pain and suffering for the child. Furthermore, I could not understand why any person who "suffered" from Achondroplasia would want her child to be similarly affected. Achondroplasia is a genetic condition that may be accompanied by considerable morbidity. Wouldn't any parent with the best interests of her child in mind want him to be "normal" and healthy? Why was it so important for her to have a child with Achondroplasia?

Listening to the speaker's story jolted me from my usual way of thinking about things. In that lecture hall, I not only confronted my own biases against disability and disease, but also began to consider to what extent my prejudices were inherent in the field of genetic counseling. Although the profession subscribes to a model of nondirectiveness, in considering my own responses, I became acutely aware of the problematic and enigmatic nature of this model.

What underlying beliefs informed my very negative response? How might this unconscious but compelling bias affect the information and advice I give to my future patients? The silence in the convention hall suggested that I was not the only person struggling with these questions. Perhaps these genetic counselors were evaluating their own responses and a professional code of ethics, which compels them to suppress those responses.

This experience was a serendipitous and beneficial occurrence for the author; but unfortunately, such experiences are not systematically available to all genetic counselors or genetic counselors in training. Without such opportunities, the beliefs and practices of the counselor with little experience of disability will most likely be influenced by the medical model of health and disease that informs the delivery of health care services in the current American medical system.

GENETIC COUNSELING AND THE MEDICAL MODEL

Historically, genetic counseling has relied upon the medical model of health and disease in creating and defining its mission. Practicing in a field that identifies risk assessment as one of its primary functions, genetic counselors "typically talk of the 'risk' of having a child with a particular genetic condition" (Davis 1997a, 7). There can be no doubt that the language used implies a negative value judgment. "Risk" is defined as "the possibility of suffering harm or loss; danger" (Morris 1985, 1121). This definition casts the event of having a child with Achondroplasia or hereditary deafness in negative terms and would explain the discomfort that genetic counselors feel in dealing with parents who wish to select *for* children with genetic conditions.

The underlying assumption here is that being born with a disabling genetic condition is a bad thing. The implied goal of medicine is to eliminate diseases, defects, and abnormalities. Based on this medical model, the implied goal of genetic counseling is to provide information that reduces the likelihood of propagating these conditions. This presents a conflict between a genetic counseling ethic that values nondirectiveness and a medical ethic that embraces eradication of disease. It also casts additional doubt upon the neutrality of the nondirective principle. Can genetic counselors truly be nondirective if they provide only negative information on the disease or condition and do not include positive viewpoints from those actually affected by it? Can genetic counselors truly be neutral if, in spite of the nondirective principle, their core bias is against abnormality? And do some members of the profession have a strong impulse against bringing an "abnormal" child into the world, even when the alternative is aborting that child?

As Adrienne Asch, a leading spokesperson of the disability movement expresses it: "Some of us are convinced of subtle and not-so-subtle bias against disabled people by some geneticists who do the science and by some genetic counselors that translate science to the public seeking help. . . . The whole genetics enterprise is permeated by the medical model of disability linking every difficulty to the physiological characteristics of the condition and not to any characteristics of the society in which people with the condition live their lives. Many of us repudiate the medical model and call for a virtual overhauling of the way genetics professionals go about their work" (1993, 3–4).

THE GENETIC COUNSELING SESSION

Genetic counseling sessions are structured to inform parents of the negative consequences of having a particular genetic condition and to make sure that they understand those consequences. By describing medical, cognitive, and psychological problems characteristic of various disorders, the prenatal genetic counselor attempts to approximate the experience of being disabled or raising a disabled child in an effort to facilitate prenatal decision-making. In these abbreviated educational sessions, counselors may discuss the pros and cons of having a genetic test such as amniocentesis, but rarely, if ever, discuss the positive aspects of having a particular condition or disease. In most cases, only detection of abnormality prompts a more comprehensive discussion, which might include viewpoints from disabled individuals and the parents who raise them.

In cases where a genetic condition is actually detected, counselors make a number of value judgments before the session even begins that may influence how a given diagnosis is perceived. For instance, in a prenatal session where testing reveals that a fetus has Down syndrome, the counselor must decide what the parents are to be told about this condition. She may present only the medical "negatives" (mental retardation, heart defects, etc.) or she may choose to include stories of parents who have raised children with Down syndrome and found it to be tremendously rewarding.

Indeed, Down syndrome presents a fruitful case to explore simply because it is one of the conditions prenatal testing most frequently detects, and evidence reveals that the manner in which the condition is reported dramatically affects the rate that such fetuses are aborted. Michael Bérubé ventures that, in general, 90 percent of fetuses diagnosed with Down syndrome are aborted (Bérubé 1996, 76). Yet, for example, the New England Medical Center reports that when women whose fetuses are diagnosed with Down syndrome are routinely introduced to families raising infants, children, and/or young adults with Down syndrome, the abortion rate is 62 percent (Parens and Asch 2000b, 8).

Such diverse writers as Rayna Rapp (1988) and Carol Gilligan (1982) report the terms in which women make reproductive decisions, revealing the web of concrete detail that shapes their responses. Without navigating the numerous philosophical shoals concerning the ethics of care, we would urge making use of the insights gleaned from the described New England practice: that moral decisions are contextual and specifically situated, made in the context of a web of relationships. The steps taken in the New England study allow prospective parents to imagine concretely what their life would be like with a child with Down syndrome and to project themselves into a textured relationship with such a child. Though genetic counselors often do follow these steps in the relatively rare cases when abnormality actually occurs, counselors do not routinely present positive images of disability in initial prenatal counseling sessions. Typically, clients attend a single informational session, undergo laboratory procedures such as blood tests and amniocentesis, and eventually hear the results by telephone when test results are normal. Thus, most clients' only experience of disability may be this initial session during which they contemplate the "reality" of being disabled or raising a disabled child, a constructed reality grounded in the scientific and medical definition of a given genetic condition or syndrome. It is in this routine, yet limited, approximation of the experience that genetic counseling may have its most profound effect on public perceptions of disability and on disabled people's lives. Specific images of living with disability might well be added to such sessions.

Detailed pictures do, in fact, profoundly affect women's views, as Nancy Press (2000) discovered in her interviews of women at a prenatal clinic concerning their views of disability. When women were questioned about their beliefs about Down syndrome, many of those interviewed evidently had been strongly influenced by the television drama *Life Goes On* (popular between 1989 and 1993) featuring an actor with Down syndrome. These women expressed such views about Down syndrome as "It's not retardedness, not exactly. . . . It looks like they have mental retardation because of their movements, but their minds are fine" or "their eyes bulge out a little; they're actually smart but their mentality just isn't like ours" (Press 2000, 227). What such research demonstrates is that perception of disabilities is highly, perhaps frighteningly, malleable, readily influenced by, in this case, one concrete dramatic presentation. Thus, prospective parents need detailed, realistic views of what life with a child with Down syndrome might be like. If counselors are to counter the claims against them, such representations should be presented not only in cases of suspected abnormality, but to all prospective parents who request or require their services.

Michael Bérubé offers a detailed portrait of raising a disabled child in his book *Life As We Know It* (1996), which he begins this way: "My little Jamie loves lists: foods, colors, animals, numbers, letters, states, classmates, parts of the body, days of the week, modes of transportation, characters who live on *Sesame Street,* and the names of people who love him" (ix). Bérubé's vivid and detailed description allows the reader initially to see Jamie, a boy, an individualized person, rather than the schematic figure of a child with Down syndrome. And as Bérubé continues with his account we see Bérubé, a highly sophisticated intellectual, passionately engaged with his son — singing, dancing, and conversing. As Bérubé describes his life with his child, he also describes his fears, his heartaches, his and his wife's grueling efforts, and the unique features of their lives that make this situation possible — flexible work schedules and full insurance. And he carefully eschews generalizations: "We think Jamie was — and is — worth every effort, of course, and we'll say so to any prospective parents who ask, regardless of how their circumstances may differ from ours. But what can we tell the Korean couple who had a child with Down syndrome during the brief time they were at school in the United States and now say they cannot return to Korea with the child because of the social stigma involved? What can we tell the mother in her forties who fears bearing a child with Down syndrome who will likely outlive her, possibly developing Alzheimer's or becoming weak and incontinent years after she's passed away?" (84–85). Reading a book such as Bérubé's and spending time with a family raising a child with Down syndrome are possible ways that prospective mothers and their partners may be able to make informed decisions about continuing a pregnancy involving a fetus that is found to have Down syndrome. According to the Down Syndrome Congress, prospective parents who learn that their fetus has a disabling trait need to receive: "a) information that seeks to dispel common misconceptions about disability and present disability from the perspective of a person with a disability, b) information on community-based services for children with disabilities and their families as well as on financial assistance programs, c) materials on special needs adoption, and d) a summary of major laws protecting the civil rights of persons with disabilities. Also people with disabilities and parents of people with disabilities should be available to talk with parents" (Parens and Asch 2000b, 36).

We would argue that what prospective parents also need are multiple, particularized accounts of how they might lead their lives with or without such a child. They need, in addition, to hear from such people as Jason Kingsley, who wrote at the age of 17: "I have a disability called Down Syndrome. My bad obstetrician said that I would never learn and send me to an institution and never see me again. No way Jose! Mom and Dad brought me home and taught me things. He never imagined that I could write a book. I will send him a copy of this book so he'll know. I will tell him that I play the violin, that I make relationships with other people. I make oil paintings. I play the piano.

I can sign. I am competing in sports, in the drama group, that I have many friends, and I have a full life" (Kingsley and Levitz 1994, 27–28).

However, genetic counselors stand as the gatekeepers to such accounts. In order to lead their clients to such information and stories, they must come to appreciate the individualized and multiplicitous nature of people with disabilities, encompassing viewpoints of disabled individuals who refute public perceptions about disability and who argue that some aspects of disability are not only constructive but desirable. They must continually work against, to employ Adrienne Asch's concept, substituting one trait of a person for the whole of that person (Asch 2000, 240). And they must identify and work against their own biases so that their introductions will not taint the process. The hope is that through these methods the gap will diminish between those who identify themselves as being without disabilities and those whom they perceive as having them. One step toward overcoming bias is for genetic counselors to appreciate fully the social construction theory of disability.

SOCIAL CONSTRUCTION THEORY OF DISABILITY

One of the recurrent themes in recent disability literature is that disability is not so much a state of being as it is a socially constructed phenomenon (Silvers, Wasserman, and Mahowald 1998, 74–75). Proponents of this theory argue that the disadvantages associated with disability stem both from the way society conceptualizes and defines disability and from a constructed environment that makes participation of disabled individuals difficult, if not impossible. In *Disability, Difference, Discrimination* (1998), Anita Silvers talks about the social model: "The social model of disability transforms the notion of 'handicapping condition' from a state of a minority of people, which disadvantages them in society, to a state of society, which disadvantages a minority of people. The social model traces the source of this minority's disadvantage to a hostile environment and treats the dysfunction attendant on (certain kinds of) impairment as artificial and remediable, not natural and immutable" (Silvers, Wasserman, and Mahowald 1998, 75). Intrinsic to this argument is the idea that society defines disability and constructs its environment relative to a standard or "norm." Though this standard changes over time and differs from culture to culture, it is usually characterized by able-bodied youth. An example of this "characterization" may be found in a 1964 article in *Lancet,* written by Dr. Marc Steinbach. In this article, a physician ponders what should be considered "normal" with regard to cardiovascular health: "There should be only one set of normals—namely, the values characteristic of young adults between 20 and 30 years" (1117).

Phillip Davis and John Bradley (1997) assert that today's standard has not changed much, except that there is an ever-increasing emphasis on perfection, and thus an ever-decreasing number of individuals who meet the ideal standard. On this continuum of normalcy, the disabled find themselves relegated to the extreme "abnormal" end of the scale. This categorizing of individuals based on an approximation of the ideal carries a concomitant implied judgment about acceptability and essential worth. Judgments about physical and mental being become global assessments of personal value and desirability. This further isolates the disabled community from society and undermines the disabled person's sense of self.

Davis and Bradley also point out that the disabled are not the only ones affected by this "ranking system." Assessing individuals relative to a standard of perfection has negative implications for all human beings, as the inability to achieve or maintain perfection is part of the human condition. At some point in each of our lives, we will all be found lacking. In "Duties and Decency" (1995), Peter Williams argues along these lines: "We all had disabilities as children; we all have them now, and we'll certainly have more of them as we get older. . . . Our stance toward people with disability is one that will apply toward us all" (98).

Thus, Williams urges us to be tolerant—not only because it is morally right to be so, but also because, in the end, it is we who will be affected by our own clemency or lack thereof. Our current attitudes towards those with disabilities will indirectly affect how society views us when the passage of time or a catastrophic event leaves us old, infirm, and "imperfect." Williams essentially argues for a variation of the golden rule, encouraging us to treat others as we will want to be treated when we are so situated. The implied message here is that seeing disabled individuals in this way will increase our awareness of our own vulnerability and of the inevitability of disability in most of our lives, and thus encourage development of a more encompassing viewpoint.

Susan Wendell (1996) takes this a step further, arguing for a "deconstruction of disability" through reconsidering both the "expectations of performance" and the "physical and social organization of societies" (37). She suggests that we address the "problem" of disability by examining the philosophical underpinnings of our attitudes toward those with disabilities, while "deconstructing" the physical barriers imposed by society. In *The Rejected Body* she reflects: "What would it mean, then in practice, to value disabilities as differences? It would certainly mean not assuming that every disability is a tragic loss or that everyone with a disability wants to be 'cured.' It would mean seeking out and respecting the knowledge and perspectives of people with disabilities. It would mean being willing to learn about and respect ways of being and forms of consciousness that are unfamiliar" (84).

Wendell envisions a world in which personal respect means not only allowing each individual the freedom to define what constitutes a meaningful existence, but also affirming that definition. That a person's perspective may be different from our own need not mean that it is valued or appreciated less. It is this view, as articulated by Wendell, that we suspect would bring genetic counselors closest to the ideal of nondirective counseling, if assimilated and applied. Some work has been done in this direction.

THE EDUCATION OF GENETIC COUNSELORS: A FEMINIST PERSPECTIVE

In her 1996 Ph.D. dissertation at Union Institute, Marsha Saxton describes her work as disability consultant for the Brandeis Genetic Counseling Program and her efforts to help genetic counseling students come to understand both the individualities of the experiences of the disabled and the political and social issues confronting those who are disabled. In her program, she wishes to foster relationships between the students, who identify themselves as nondisabled, and persons in the community who identify themselves as disabled. In her research design, she is heavily influenced by a feminist methodology that privileges relationship. She finds this particularly valuable given her subject because, as Saxton observes, one of the greatest societal problems those with disabilities confront is their objectification (1996, 47).

To foster a relationship of equality between the students and members of the disability community, Saxton disrupts the usual power relationships that obtain between medical professionals and those with disabilities. Members of the community were given the title of "disability consultant." Saxton asked students to learn of their interviewees' lives and perspectives by interviewing them three times over the course of a semester. Saxton interviewed both the students and consultants before and after their experiences. In keeping with her philosophical and methodological assumptions, Saxton included their remarks unmediated by herself as researcher so that the reader might develop the analogue of a relationship with the participants. One can hear the students changing. They approach their first interviews self-consciously, painfully aware of their own attitudes of pity and fear. But although these attitudes do not completely disappear, their attitudes shift, as evidenced by Alisa, one of the students quoted by Saxton: "For me, it was a great experience. I really had a shift in the way I saw things. . . . The person that I interviewed is so alive and vibrant. I remember one time leaving his home and getting into the car and turning on the radio and saying, 'Yes, that's [becoming disabled] not such a big bad thing to happen.' I'm driving along and I'm moving my body and then I think, 'Oh my god, would I like dancing!' It really hit me that I got into his world, I saw so many other things!" (1996, 133).

In reading Saxton's study, one is reminded of the feminist methodology advocated by Maria Lugones and Elizabeth Spelman in their much-anthologized article, "Have I Got a Theory for You!" (1983). In the article, two voices are heard, unmuted, one Hispanic, one Anglo, each addressing the other from her own experience, her own perspective. Underlying the article as a whole is Lugones' assertion to non-Hispanic feminists: "The only motive that makes sense to me for your joining us in this investigation is the motive of friendship, out of friendship. A nonimperialist feminism requires that you make a real space for our articulating, interpreting. . . . I see the 'out of friendship' as the only sensical motivation, . . . because the task for you is one of extraordinary difficulty" (Lugones and Spelman 1983, 577). Such a technique is the one employed by Saxton. If the students and consultants initially come together out of something more artificial than friendship, Saxton creates an atmosphere in which friendships can develop. Presumably, what happens to students in such a program is that they learn not only something about the experiences of those with disabilities but also something invaluable about themselves, and about the constructed and privileged position they occupy as able-bodied persons. It is something like what bell hooks (1995) describes when she observes "how white people who shift locations . . . begin to see the world differently" (1995, 49). Programs like Saxton's should be incorporated into genetic counseling programs. Certainly, such steps will not end the difficulties, just as Lugones and Spelman coming together to listen to each other does not end oppressive ethnic and racial relationships within feminism or within the society as a whole. However, such steps would lead in the right direction.

A CAUTIONARY NOTE

As salutary as Saxton's methods are, we must remember collectively and individually how difficult it is to alter our own perspective. A personal narrative and a conversation between mother and son remind us.

Deborah Kent in "Somewhere a Mockingbird" (2000) writes of the chasm between herself and her beloved husband, and between herself and her devoted parents, regarding her blindness, a gulf of which she was unaware until she and her husband contemplated having children. Her short personal narrative stands as an agonized illustration of the inevitable gap between the able-bodied and those with disabilities, even in the most loving and intimate relationships.

Early in her essay, Kent explains that because her brother is also blind, there is reason to suspect a genetic basis for her blindness. She sketches her own attitude toward her blindness growing up, mostly in terms of the prejudiced attitudes it engendered. But she explains: "I premised my life on the conviction that blindness was a neutral characteristic. It created some inconveniences, such as not being able to read print or drive a car. . . . But in the long run, I believed that my life could not have turned out any better if I had been fully sighted. If my child were blind, I would try to ensure it every chance to become a self-fulfilled, contributing member of society." Her husband says that he agrees with her completely. But later, he reveals that if their child were born blind, "I'd be devastated at first, but I'd get over it" (58).

Kent's understanding of her husband's remark is that he is prejudiced against blindness, that although his love for her has provided an exception, his bias itself remains intact. One might see her interpretation of her husband's position analogously with racial prejudice. Many have written about how the racist feels more comfortable if he can claim that there are acceptable blacks, blacks who are his friends, while his basic prejudice remains untouched. Adrienne Rich (1986) describes "tokenism" in this way, offering as an example someone who accepts a woman as a member of a given discipline, but who nevertheless allows all the institutional biases against "women's ways of thinking" to remain unaltered (3–4). Kent (2000) writes: "What I understood was that Dick, like my parents, was the product of a society that views blindness, and all disability, as fundamentally undesirable. All his life he had been assailed by images of blind people who were helpless, useless, and unattractive, misfits in a sight-oriented world. I had managed to live down that image. Dick had discovered that I had something of value to offer. But I had failed to convince him that it is really okay to be blind" (58–59).

She later describes her husband's and her parents' celebration when they discover that the infant she gives birth to is sighted, a celebration in which she cannot participate emotionally. She does take solace in her relationship with her husband: "But I recognize that people can and do reach out, past centuries of prejudice and fear, to forge bonds of love. It is a truth to marvel at, and a cause for hope and perhaps some small rejoicing" (62).

But the predominant tone of Kent's essay is one of pain and loneliness; it conveys a sense that even with her beloved husband she cannot share her sense of who she is. But within the essay are some notes of hope as well, such as in the song of the titular mockingbird that "sang so boldly in a place where no one thought it belonged" and the voice of the physician with whom the couple had consulted. He comments, "You have a good life, don't you? . . . Go home and have a dozen kids if you want to!" (60). Thus, Kent's essay sends a double message: that it is nearly impossible for someone to transcend his situated view even within the most loving and intimate of relationships, but that on rare occasions, someone such as this physician does, and such an event is miraculous.

An article written by Eva Kittay and her adult son, Leo, (2000) makes a related but different point. They write about Eva Kittay's daughter Sesha, Leo's sister, who has profound disabilities. As mother and son address each other, their deep love for Sesha, their great joy in her existence, permeates every sentence. They have lived a lifetime together in this loving family, yet evidently they had not, until the occasion of writing this article, communicated with each other about what the termination of a pregnancy of a fetus with a condition such as Sesha's would express. Leo speaks from the position of a son about what aborting such a fetus might mean to him: "The love my parents have for me is a condition of my being mentally and physically sound, not just of being a child of theirs. Rephrasing this: The only reason my parents want me is that I'm relatively smart and fit" (169).

And Eva Kittay, while lingering over Sesha's incredible sweetness, also talks, as she addresses Leo, of her maternal pain as she ponders the woman Sesha might have been: "With tears in my eyes, I think about the young woman of twenty-seven who might be a graduate student like my wonderful graduate students, or be thinking about marriage or be out on the ski slopes with you" (190). The article exemplifies communication at its best: two thoughtful, loving people in an intimate relationship addressing each other on an emotional and intellectual level. But beyond their philosophical disagreements, there is a gap between them that never closes, for one is the brother and the other the mother of a woman with disabilities. And they derive different meanings from their respective experiences.

Perhaps there is no more powerful demonstration, for both the need for communication and the need to hear multiple voices. The experience of families of persons with disabilities is a multifarious one; understanding comes as the result of great effort. The conversation of mother and son can tell us a great deal about listening with care, and perhaps about that which can never be communicated. Feminism asks us to hold in mind both the inevitable gaps between us and the communicative approaches to closing those gaps.

NOTES

This paper is based, in part, on Patterson, Annette. 2000. Genetic counseling and the disabled; Re-evaluating the connection. Master's thesis, Sarah Lawrence College. She wishes to thank Laura Hercher for her editorial expertise, and Caroline Lieber, director of the Sarah Lawrence counseling program, for her advice and continued support of this work.

Martha Satz would like to acknowledge the inspiration of her student, Robbie Van Schoick, for both her wise words and for her persistence under adversity.

1. See Bosk 1992; Rapp 1988; Suter 1998; and Kessler 1997a and 1997b for the problematics of nondirective counseling. Some disturbing evidence of the possible biases of genetic counselors appears in Helm, Miranda, and Chedd 1998; Marteau, Slack, Kidd, and Shaw 1992; and Wertz 2000.

2. We should note that some disability activists argue that there is an intrinsically negative component to disabilities. Adrienne Asch says, "The inability to move without mechanical aid, to see, to hear, or to learn is not inherently neutral. Disability itself limits some options. . . . It is not irrational to hope that children and adults will live as long as possible without health problems or diminished capacities" (1989, 73).

3. For a sense of the richness of the Hassidic life, see the revealing 1997 film *A Life Apart*.

4. Feinberg himself realizes that this is a severely restricted notion but believes it can be narrowly maintained.

REFERENCES

Asch, Adrienne. 1989. Reproductive technology and disability. In *Reproductive laws for the 1990s,* ed. Sherrill Cohen and Nadine Taub. Clifton, N.J.: Humana Press.
———. 1993. The human genome and disability rights: Thoughts for researchers and advocates. *Disability Studies Quarterly* 13 (3): 3–5.
———. 2000. Why I haven't changed my mind about prenatal diagnosis: Reflections and refinements. In *Prenatal testing and disability rights,* ed. Erik Parens and Adrienne Asch. Washington, D.C.: Georgetown University Press.
Bérubé, Michael. 1996. *Life as we know it: A father, a family, and an exceptional child.* New York: Random House.
Bosk, Charles. 1992. *All God's mistakes: Genetic counseling in a pediatric hospital.* Chicago: University of Chicago Press.
Collins, Patricia Hill. 1990. *Black feminist thought: Knowledge, consciousness, and the politics of empowerment.* Boston: Unwin Hyman.
Crouch, Robert. 1997. Letting the deaf be deaf: Reconsidering the use of cochlear implants in prelingually deaf children. *Hastings Center Report* 27 (4): 14–21.
Davis, Dena S. 1997a. Cochlear implants and the claims of culture? A response to Lane and Grodin. *Kennedy Institute of Ethics Journal* (3): 253–58.
———. 1997b. Genetic dilemmas and the child's right to an open future. *Hastings Center Report* 27 (2): 7–15.
Davis, Phillip, and John Bradley. 1997. The meaning of normal. *Perspectives in Biology and Medicine* 40 (1): 68–77.
Feinberg, Joel. 1991. The child's right to an open future. In *Freedom and fulfillment.* Princeton: Princeton University Press.

Felker, Kitty S. 1994. Controlling the population: Views of medicine and mothers. *Research in Sociology of Health Care* 11:25–38.

Gilligan, Carol. 1982. *In a different voice: Psychological theory and women's development.* Cambridge, Mass.: Harvard University Press.

Harding, Sandra. 1993. Rethinking standpoint epistemology: What is strong objectivity? In *Feminist epistemologies,* ed. Linda Alcoff and Elizabeth Potter. New York and London: Routledge.

Hartsock, Nancy. 1983. The feminist standpoint: Developing the ground for a specifically feminist materialism. In *Discovering reality,* ed. Sandra Harding and Merrill Hintikka. Dordrecht: Reidel.

Helm, David T., Sara Miranda, and Naomi Agoff Chedd. 1998. Prenatal diagnosis of Down syndrome: Mothers' reflections on supports needed from diagnosis to birth. *Mental Retardation* 36 (1): 55–61.

hooks, bell. 1995. *Killing rage: Ending racism.* New York: Henry Holt and Company.

James, William. 1897. The will to believe. In *The will to believe, and other essays in popular philosophy.* New York: Longmans, Green, and Co.

Kass, Leon. 1983. Implications of the human right to life. In *Intervention and reflection: Basic issues in medical ethics,* ed. Ronald Munson. Belmont, Calif.: Wadsworth.

Kent, Deborah. 2000. Somewhere a mockingbird. In *Prenatal testing and disability rights,* ed. Erik Parens and Adrienne Asch. Washington, D.C.: Georgetown University Press.

Kessler, Seymour. 1997a. Genetic counseling is directive? Look again. *American Journal of Human Genetics* 48 (3): 466–67.

———. 1997b. Psychological aspects of Genetic Counseling XI: Nondirectiveness revisited. *American Journal of Medical Genetics* 72 (2): 164–72.

Kingsley, Jason, and Mitchell Levitz. 1994. *Count us in: Growing up with Down syndrome.* New York: Harcourt Brace.

Kittay, Eva Feder, with Leo Kittay. 2000. On the expressivity and ethics of selective abortion for disability: Conversations with my son. In *Prenatal testing and disability rights,* ed. Erik Parens and Adrienne Asch. Washington, D.C.: Georgetown University Press.

A life apart: Hasidism in America. 1997. Produced and directed by Menachem Daum and Oren Rudavsky. 95 min. First Run/Icarus Films. Videocassette.

Lugones, Maria C., and Elizabeth V. Spelman. 1983. Have we got a theory for you! Feminist theory, cultural imperialism and the demand for "the woman's voice." *Women's Studies International Forum* 6 (6): 573–81.

Marteau, T. M., J. Slack, J. Kidd, and R. W. Shaw. 1992. Presenting a routine screening test in antenatal care: Practice observed. *Public Health* 106 (2): 131–41.

Morris, Jenny. 1991. *Pride against prejudice: Transforming attitudes to disability.* Philadelphia: New Society.

Morris, William, ed. 1985. *The American heritage dictionary of the English language.* New York: Houghton Mifflin.

Noddings, Nel. 1984. *Caring, a feminine approach to ethics and moral education.* Berkeley: University of California Press.

Parens, Erik, and Adrienne Asch. 2000a. Introduction. In *Prenatal testing and disability rights,* ed. Erik Parens and Adrienne Asch. Washington, D.C.: Georgetown University Press.

———. 2000b. The disability rights critique of prenatal genetic testing: Reflections and recommendations. In *Prenatal testing and disability rights,* ed. Erik Parens and Adrienne Asch. Washington, D.C.: Georgetown University Press.

Press, Nancy. 2000. Assessing the expressive character of prenatal testing: The choices made or the choices made available. In *Prenatal testing and disability rights,* ed. Erik Parens and Adrienne Asch. Washington, D.C.: Georgetown University Press.

Press, Nancy Anne, and Carole H. Browner. 1994. Collective silences, collective fictions: How prenatal diagnostic testing became part of routine prenatal care. In *Women & prenatal testing: Facing the challenges* of *genetic technology,* ed. Karen H. Rothenberg and Elizabeth J. Thomson. Columbus: Ohio State University Press.

Rapp, Rayna. 1988. The power of positive diagnosis: Medical and maternal discourses on amniocentesis. In *Childbirth in America: Anthropological perspectives,* ed. Karen L. Michaelson. South Hadley, Mass.: Bergin and Garvey.

Rich, Adrienne. 1986. What does a woman need to know? In *Blood, bread, and poetry: Selected Prose, 1979–1985.* New York, London: W. W. Norton and Company.

Ruddick, William. 2000. Ways to limit prenatal testing. In *Prenatal testing and disability rights,* ed. Erik Parens and Adrienne Asch. Washington, D.C.: Georgetown University Press.

Sacks, Oliver W. 1989. *Seeing voices: A journey into the world of the deaf.* Berkeley: University of California Press.

Saxton, Marsha. 1984. Born and unborn. In *Test-tube women: What future for motherhood,* ed. Rita Arditti, Renatti Duelli Klein, and Shelley Minden. Boston: Routledge/ Kegan.

———. 1996. Disability feminism meets DNA: A study of an educational model for genetic counseling students on the social and ethical issues of abortion. Ph.D. diss., Union Institute.

———. 2000. Why members of the disability community oppose prenatal diagnosis and selective abortion. In *Prenatal testing and disability rights,* ed. Erik Parens and Adrienne Asch. Washington, D.C.: Georgetown University Press.

Silvers, Anita, Dave Wasserman, and Mary Mahowald. 1998. *Disability, difference, discrimination.* Lanham, Md.: Rowman and Littlefield.

Steinbach, Marc. 1964. The normal in cardiovascular diseases. *Lancet* 2:1116–18.

Suter, Sonia M. 1998. Value neutrality and nondirectiveness: Comments on future directions in genetic counseling. *Kennedy Institute of Ethics Journal* 8 (2): 161–63.

Tollifson, Joan. 1997. Imperfection is a beautiful thing. In *Staring back: The disability experience from the inside out,* ed. Kenny Fries. New York: Plume.

Walker, Ann Platt. 1998. The practice of genetic counseling. In *A guide to genetic Counseling,* ed. Diane L. Baker, Jane Schuette, and Wendy Uhlmann. New York: Wiley-Liss.

Wendell, Susan. 1996. *The rejected body.* New York: Routledge.

Wertz, Dorothy C. 2000. Drawing lines: Notes for policymakers. In *Prenatal testing and disability rights,* ed. Erik Parens and Adrienne Asch. Washington, D.C.: Georgetown University Press.

Williams, Peter. 1995. Duties and decency. *The Mount Sinai Journal of Medicine* 62 (2): 98–101.

4

The Natural Father: Genetic Paternity Testing, Marriage, and Fatherhood

Gregory E. Kaebnick

The emerging phenomenon of genetic paternity testing shows how good science and useful social reform can run off the rails. Genetic paternity testing enables us to sort out, in a transparent and decisive way, the age-old but traditionally never-quite-answerable question of whether a child is genetically related to the husband of the child's mother. Given the impossibility of settling this question for certain, British and American law has long held that a biological relationship must almost always be assumed to exist. According to what is known as the "marital presumption" or "presumption of legitimacy," a child born to a woman within a marital relationship is assumed to be the biological child of the woman's husband unless he was absent, impotent, or sterile. In other words, if paternity was not a physical impossibility for the husband, there was a nearly irrebuttable presumption that he was the father of the child.[1] The husband was locked into the role of fatherhood.

Some now argue that this presumption is outmoded.[2] Science makes certainty possible, and family members should be able to use this science to gain knowledge about their true family relationships, to extricate themselves from relationships that have been scientifically undermined, and to establish new relationships on the basis of the scientific evidence. This position now permeates the media[3] and has given birth to a mini-industry.[4] Additionally, it is beginning to wrest American law away from its traditional presumptions: an increasing number of states are introducing and approving legislation that makes it easier to use genetic paternity testing in court.[5] As the testing establishes itself in society, divorce lawyers could end up recommending that their clients obtain genetic paternity testing.[6]

The pressing conceptual issue behind paternity testing is the nature of parenthood. Paternity testing encourages us to suppose that a parental relationship to a child is fundamentally a genetic relationship, or at least necessarily includes a genetic relationship. Paternity testing also encourages us to suppose that a parental relationship can be reassessed years or decades after the parties in the relationship thought it was established.

Insofar as paternity testing enables people to reopen the question of parenthood, it enjoys some support from a broad and ongoing social transformation. In the mid-twentieth century, the underlying commitment behind much of family law was preservation of the social order. Family law enforced social conventions about the familial roles people filled—as parent, husband, wife, or child. Increasingly, as the legal historian Michael Grossberg has argued, this commitment gave way to a "commitment to individual choice and private ordering" (p. 7).[7] People enjoyed greater freedom both to enter into and get out of marriages, and their rights and responsibilities within families were reassessed and, in some measure, made more equitable. Perhaps the best example of this transformation is the emergence of no-fault divorce, but rules about whom one may marry and cohabit with, about child custody, about marital rape, about birth control and abortion, and about children's rights were also affected.[8]

In almost every case, this broad reform is desperately needed and still unfinished, but I will argue against unrestricted genetic paternity testing. The central conceptual issue raised by paternity testing is the nature of parenthood. My general conclusion—I will not set out policy options in detail—is that society may reasonably constrain individuals' use of and access to DNA-based paternity testing—at least if those individuals have had the opportunity to develop relationships with the children and they seek the testing to reassess the existing family relationships. I approach this issue by setting out the sorts of cases that involve paternity testing and commenting along the way on the range of issues involved in them.

A TYPOLOGY OF CASES

Sometimes, paternity testing is an inadvertent side effect of medical tests. In one of the better publicized legal cases involving claims about biological paternity, for example, the father discovered that he was not the biological father of a child he was helping raise when he learned that the child had cystic fibrosis but that he was not a carrier for the disease.[9] Cases of accidental discovery are more likely to feature completely unsuspecting men who are also actively involved as fathers, and therefore are somewhat more likely to raise questions about whether the man has been betrayed by the mother. These are the classic "duped dads," as they have sometimes been dubbed by the media—exact opposites of the uninvolved yet biologically related dead-beat dads.

In other cases, paternity testing is intentional. Commentators frequently organize these cases into rough categories according to what the testing is intended to establish. There are four primary categories. In one, genetic paternity testing is sought by a man to terminate parental responsibilities.[10] Frequently, this use of paternity testing is but one feature of the dissolution of the entire family. Test results showing no genetic relationship may prompt the man to terminate the marriage as well as the parental relationship; or the genetic test may be sought in the midst of divorce proceedings. Typically, the man argues that if he is not genetically related to the child, then he should not be obligated to support the child financially.

In a second and very common type of case, paternity testing can be sought by the child's mother to *impose* parental responsibilities—chiefly, the responsibility to help support the child financially. She may also be under pressure from the state to locate the genetic father, because if no man is located to provide child support payments, the financial burden may shift to the state itself. Alternatively, the state itself may require the testing. Federal welfare policy over the last several decades has given increasing support to the use of paternity testing to enforce child support, culminating in the *Personal Responsibility and Work Opportunity Act* of 1996, which required states to give child-support enforcement agencies authority to order genetic testing.[11]

The other two kinds of cases involving possible parents concern the establishment of parental rights. Paternity testing can be sought by a man to *assert* parental rights, possibly over against the mother, or possibly over against a presumptive father.[12] It can also be sought by a mother to *deny* parental rights. As in the first category, genetic testing sought by the mother to sever parental rights is commonly associated with divorce proceedings.[13]

To these four categories must be added an assortment of other types of cases in which genetic paternity testing might be sought by people other than the presumptive or putative parents. Paternity testing can be sought indirectly by other relatives, through tests to determine whether someone is a genetic grandparent, full sibling, aunt, or uncle.[14] It could also be sought by children, with a variety of possible motivations: to discover "true parentage," as occurred when Sally Hemmings's descendants sought posthumous testing to affirm that they were descended from Thomas Jefferson, to establish lines of inheritance, to establish a medical history, or merely out of curiosity. (How many adolescents have hoped that they were not biologically related to their parents?)

Individuals' efforts to gain control over their familial relationships are illustrated in these categories to various degrees. The most straightforward

instances are those cases in the first and fourth categories, in which paternity testing is associated with divorce proceedings. Indeed, it is likely that,
in some instances, the motivation to procure testing is not genuinely to discover whether a parental relationship exists but to bring about an outcome
that the individual wants anyway, whatever the biological facts may be.

In all of these types of cases, questions about the man's relationship to the
child tend to get mixed up with questions about his relationship to his wife.
In particular, we are likely to be distracted by the thought that the child is the
result of a sexual betrayal. If we are, then we will translate decisions about
the man's relationship with the child into contests between adults about their
relationship with each other, and specifically about how they will assign
parental rights and responsibilities with respect to a now somewhat-invisible
child. But sexual betrayal is not, after all, itself a feature of the relationship
between a man and the child. It is a feature of his relationship with his wife.
Even if the parental relationship is construed merely as a biological phenomenon, the marital sexual betrayal can be only *connected* to the facts of parentage.

THE NATURAL FATHER

Because our starting typology organizes cases specifically in terms of types
of contests between adults, it does not raise the question that we most need to
attend to: How should we understand the relationship between the man and
the child? What, in other words, is "fatherhood"?

The main premises of the argument for paternity testing are that parenthood is a biological phenomenon and that determinations of parentage should
therefore adhere to the biological facts. For most of the course of Western history, a biological relationship has indeed been accepted as the overriding
criterion of fatherhood. Further, this claim has often been cast as a matter of
natural law. In an older and particularly stark form of this position, a child
was held to *belong,* as property, to the man who had sired it—in nearly the
way that any artifact would presumptively belong to the person who manufactured it—unless of course that man himself was the property of another.[15]
Weaker forms of the position continue to enjoy support both with the public
and in court. Starting from the premise that parenthood is biological, for example, one might argue that a parent is to be regarded not as owner but merely
as steward of the child—that the central moral feature of the parent-child relationship is the parent's responsibility to protect and advance the child's
interests, and the parent's authority over the child is derivative from that responsibility. Proponents of such a position claim that there is a natural bond

between parent and child that makes the parent the ideal person to raise the child. Thus Samuel Pufendorf wrote: "Within the family, nature creates the link between the social order and parental rights. Nature works on parents to 'stir up their Diligence, wisely implant[ing] in them a most tender Affection towards these little Pictures of themselves.'"[16]

A recent variant of this position draws from contemporary views about the genes' role as the "blueprint" for the person. True family bonds are encoded directly in genetic relationships. The genes so thoroughly control a person's nature, according to this position, that two people who are genetically related to each other will immediately feel a personal bond between them. In Kaja Finkler's study of adopted children who have searched for their genetic parents, a young woman explains that "with your own you instinctively feel a kinship connection that passes through. . . . *It's a memory code in the genes.*"[17] Those who never recover genetically underpinned relationships—those who are adopted, for example—will experience "a hole at the center of their being." They are cut off from their "soul mates."[18]

This reductive and deterministic "view of the genes" contribution to a person's character is increasingly popular in the media, as anyone who spends much time in grocery store check-out lines can attest to, but genetic scientists now doubt that the genes exert the kind of control that this latter picture imputes to them. Human behavioral traits—those aspects of a person's character on which an interpersonal bond would depend—are thought to depend on extremely complex interactions both among genes themselves and among genes and the physical and social aspects of environment.[19] Such views as Finkler quotes probably derive from, more than support, the popular view of genetic science.

Even Pufendorf's more moderate explanation for why a genetic relationship is the natural basis for a parental relationship has come into question, however. There are no studies of the impact on children's welfare of a discovery of misattributed paternity—much less of the impact of undiscovered misattributed paternity!—but studies of analogous groups are reassuring. The closest analog is provided by children conceived through sperm donation, as they are genetically related to their mother but not to their rearing father and they have a social relationship with both beginning at conception. In her review of studies on children's psychological development, Susan Golombok concludes that these children tend to do as well as children conceived naturally, and she tentatively infers that a genetic relationship to the rearing father provides no apparent benefit to a child.[20]

Adopted children offer another analog, though a more distant one, as they are genetically related to neither parent and their social relationship begins sometime after birth. Here, the news is cautionary at first blush: behavioral problems are more likely to be diagnosed in adopted children than in children who are genetically related to their parents.[21] Yet the lack of a genetic parental relationship may not be the cause of the gap.[22] Perhaps more behavioral problems are diagnosed in adopted children simply because rearing parents and others are looking harder for problems. Alternatively, adopted children might have more behavioral problems to diagnose, but for reasons that shed no light on the likely welfare of children involved in cases of misattributed paternity. Mothers who put their children up for adoption are more likely than others to have experienced a variety of medical and psychological problems during pregnancy and childbirth, any of which might have harmed the child. Additionally, the child's experiences before the adoption and of the adoption process itself might have an impact on the child's well-being; children adopted as infants fare much better than children adopted later in childhood. The mere fact of being adopted also has social and psychological repercussions: because adopting a child is often seen as a less desirable way of becoming a parent, the children may be stigmatized by peers and may feel unwanted by their parents—both by their biological parents and by their adoptive parents.

THE REARING FATHER

There might, of course, be times when it is indeed in a child's interest to remain with its genetic parents. But if we do not suppose that children *belong* to those who brought them into the world, nor that children should be raised by their biological parents as a matter of natural law, nor that human nature is so determined by the genes that genetically related people will have a personal relationship by a kind of physical necessity, then we will recognize that the parental relationship, like all human relationships, is first and foremost a psychological and social phenomenon and may or may not be entwined with an underlying biological relationship. This view is sometimes known as the functional or psychological account of parenthood. It is described by Joseph Goldstein, Anna Freud, and Albert Solnit, in an influential work on child development, as follows: "Whether any adult becomes the psychological parent of a child is based on day-to-day interaction, companionship, and shared experiences. The role can be filled either by a biological parent or by an adoptive or by any other caring adult."[23] In his extended discussion of the meaning of parenthood, Tom Murray has suggested the somewhat less clinical and more humanistic term "rearing parent" for this conception.[24]

The notion of rearing parenthood has ancient roots, traceable at least to Roman conventions and law concerning parent-child relationships.[25] The concept also lies hidden within the marital presumption. That parenthood *should* involve a biological relationship is plainly presupposed by the marital presumption. However, a recognition that a family is something humans establish, not something that biological relationship establishes for them, is what makes the presumption workable. Plainly, the presumption can fly in the face of the facts, even of the very widely known facts; all the same, implies the presumption of legitimacy, a family relationship is possible. This fact allows the presumption to become, as Janet Dolgin has written, "a substitute for, rather than a presumption about, some underlying biological reality."[26] In effect, the very fact that the marital presumption was all but irrebuttable shows that a biological relationship was not taken to be a necessary condition of parenthood.

To refine the typology of cases with which the discussion began, we need to know whether the man whose paternity is in question is the child's *rearing father*. We need to know whether he has been supporting and raising a child in the way that fathers do and whether the child has been looking to him as to a father. An individual who has been successfully filling this role is the best candidate, on the available evidence, to continue to support the child, to uphold and advance the child's interests. Furthermore, if parenthood is a fundamentally social and psychological phenomenon, then the central value of parental relationship is precisely that it upholds and advances the child's interests and, secondarily, the interests of society and of the adults whose lives are invested in that child—that is, of the rearing parents. The child's interests are foremost among these three because of the asymmetry of the parental relationship: the child, who is uniquely dependent on the adults raising him, simply has more at stake in the parental relationship than do the adults.[27]

Given these considerations, the results of paternity testing will be rendered moot in some kinds of cases. First, it will often be inappropriate to allow a man who has an established parental relationship with a child to be completely released from it. Indeed, this appears to be a reasonable presumptive rule.[28] It is not a presumption of *legitimacy,* because it makes no claims about biology. Nor is it the *marital* presumption of old, because it is grounded in the social role that the man has assumed toward the child, not in his role toward the child's mother. Its justification is that the man has effectively assumed parental obligations, by representing himself both to the child and to the world as the child's parent and thereby encouraging the child to become dependent on him and perhaps making it less likely that any other man could take a parental role. Having assumed that role, it is in most cases much in the best interests of the child that the relationship continue, in some form. Of course, it would be counterproductive to adopt a social policy that forces the man to persevere in every aspect of a parental role—only a financial obligation can be maintained by force—but according to this line of reasoning, it seems that the man is still under a *moral* obligation to continue more than a merely financial relationship.

Second, it will often be inappropriate to *deprive* him of the parental relationship—the goal of those cases in the third and fourth categories, in which a woman or another man seeks paternity testing to reassign parental rights. This appears to generate another presumptive rule. The parental rights ride along with the relationship and properly belong, we may presume, to whatever person has been building and perpetuating that relationship. It will probably be in the interests of the child that the relationship continue. And it will probably be in the interests of the man, who has by now invested himself psychologically in the relationship.

THE CLAIMS OF BIOLOGY

I have argued in favor of viewing parenthood as a psychological instead of a genetic relationship, and decisions about paternity as psychosocial rather than genetic questions. But there must be something to say on behalf of genetic relationships. If a genetic relationship were of no importance at all, then there would be no good objection to adopting a policy that involved randomly redistributing babies in the hospital. But such a proposal is unsettling, if not appalling. Plainly, a genetic relationship can also have some importance in parenthood.

One way in which the genetic relationship is important is that it is a feature of a causal relationship. Typically, an adult is the genetic parent of a child only when the adult has knowingly acted in such a way that a child could be created. We often hold each other responsible for the foreseeable consequences of our actions, and given our current social mechanisms for rearing and supporting children, it is reasonable to hold people accountable when through their actions they bring a person into existence. Here we have another presumptive rule: in the context of our current social institutions for raising children, one way of undertaking a responsibility to help cover the material costs of raising a child is to have a child.

Few will argue with this rule. The trick is to delimit it. Fathers' rights advocates, for example, tend to accept it but then use it to argue, first, that genetic fathers have not only parental responsibilities toward their child but rights to the child and, second, that men who are shown not to be genetic fathers of a child have neither rights nor responsibilities toward a child. Evidence that is sufficient to "convict" a man of parenthood ought by parity of reasoning to be sufficient to uphold a man's claims to parenthood or to "exonerate" him of it. The argument is attractively simple, but the response is by now familiar. If parenthood is (like all other human relationships) first and foremost a social and psychological phenomenon—in this case a *rearing* relationship—then either genetic testing never establishes parenthood at all or it establishes only one biological feature that is commonly part of the social relationship. Or, we might say, it establishes one, narrow, biological *kind* of parenthood—not a relationship between persons, but one between organisms. However we put it, what establishes the biological and causal relationship can in no way be assumed to establish a personal relationship (and its attendant rights) nor to disestablish a personal relationship (and its attendant responsibilities).

The genetic relationship is, of course, commonly also one aspect of the social and psychological relationship, and this gives us much more to say about its importance. A genetic relationship can often offer a variety of social and psychological benefits. The most prosaic of these is that a transparent biological relationship gives children easier access to their genetic relations' medical histories, and so better knowledge of their own medical risks. The child also may gain access to various social benefits, ranging from material benefits that are conditional on genetic relatedness to less tangible social and psychological benefits, such as those enjoyed by people who can prove that they are descendants of Thomas Jefferson.

A variety of other possible benefits are conditional in some measure on contemporary views about parenthood and biology. For example, again given contemporary views about parenthood and biology, a biological relationship sometimes contributes toward a psychological relationship. It can provide one way in which the range of experiences that the child and its parents share can be extended, the story that the child and its parents tell about themselves lengthened and deepened. It means that the caring and nurturing begin before the child has even quite appeared on the scene. It gives parents the very earliest possible experiences one can have of another person—the prenatal stirrings and sonographic images that we search (however foolishly) for clues to the child's emerging personality—and it enables parents and children to draw connections (however groundlessly) between a child's traits and those of parents and other relatives.

Knowledge about the biological relationship can also touch on the child's and the adults' sense of self. Partly by discovering shared traits that may be traceable to shared genetic factors, partly because of common sentiments about the value of genetic relatedness, a genetic relationship perhaps sometimes helps children develop a sense of identity. Adopted children who do not know the identity of their genetic relatives occasionally claim that they lack a sense of their own identity, and they sometimes launch exhaustive searches to locate their genetic parents for no other reason than to see where they fit into the genetic web of humankind, believing that this understanding will give them a better sense of who they are.[29]

For all of these reasons and for others besides, a genetic parental relationship can be one of the things that gives meaning to life. For the parent, it can be meaningful to participate in the project of physically bringing a child into the world. For the child, it can be meaningful to feel that one's physical existence is owed partly to your parents' own meaning-making. Being situated in this way in another person's story can feel good.

Many of these points are suggested by James Lindemann Nelson's concept of the "genetic narrative" view of the family. Explains Nelson:

> The role of genetic connections in our lives can be seen, not as "gene calling to gene," but rather as a part of our interest in perceiving the connections between our lives and the lives of others—connections which add depth and richness to the continuing story in which we participate, and which can therefore be referred to as narrative connections. (p. 81)[30]

These claims about the benefits of a genetic relationship are all either limited or provisional. They do not generate a manifesto for the Western family. They do not establish that parenting that involves a biological relationship is intrinsically better or more meaningful than other sorts of parental relationships, which may offer their own special benefits. There are other ways of obtaining knowledge of one's genetically acquired medical risks than by being raised by one's genetic parents, nor does being raised by one's genetic parents guarantee that one knows those risks. The meaning, the sense of identity, and the richness of shared life-story to which genetic parenthood can contribute are themselves partly conditional on Western views of the family, not intrinsic features of genetic parenthood and they would be more or less canceled out if the genetic parents did not seize the moment at the child's birth to establish a rearing relationship.

The psychosocial benefits also do not establish that men who are genetically related to their children will tend to be better rearing parents. That last claim is already undermined by studies of children conceived with donor-provided sperm, but even if it were true, if we regard parenthood as a social and psychological relationship, we would tend to look first at the existing relationships. If one man already is functioning well as a parent, we would tend to favor preserving that relationship, particularly because disrupting a parent-child relationship appears to be detrimental to a child's health.

Thus these considerations would count for very little in most cases involving genetic paternity testing, particularly when the testing is sought to test the paternity of a man who has already been raising a child as his own for some time.

HOW GENETIC TESTING MIGHT BE LIMITED

If these observations are plausible, then it would be appropriate to impose a variety of constraints on genetic testing when it is sought for purposes of re-assessing and reordering existing parental relationships. These constraints could take different forms.[31]

First, constraints might bear on how the testing is used. When the results of genetic testing are intended for use in legal paternity cases, for example, it would be appropriate to limit their weight and perhaps even to prevent them from being introduced as evidence in the first place. They will have weight chiefly in proceedings initiated by the mother to force a biological father to provide at least financial support to a child. In hearings to decide whether a rearing father has ongoing parental rights and responsibilities toward a child, they should typically either have very little weight or not be admissible as evidence.

Because of the traditional strength of the marital presumption, nearly all jurisdictions barred the results of genetic tests in paternity cases in which a presumptive father sought the test to challenge his paternity—at least until the recent wave of state legislation that makes it easier to use genetic testing, either by loosening the evidentiary restrictions on genetic paternity testing or by increasing the weight that knowledge of genetic relations should be given.[32] Both these new laws and the traditional restrictions that they seek to roll back explicitly assume the mistaken genetic view of parenthood (even though the old restrictions implicitly affirm that parenting is not fundamentally genetic). If parenthood is fundamentally a psychological and social phenomenon, the reason for blocking use of genetic paternity testing to decide whether the rearing father is the legal father is not any presumption about the biological facts; the reason is that, if we know who has filled the role of rearing father, we already know enough. The test for genetic paternity simply does not address the most important consideration.

Oddly, even some of those who object to the new wave of testing-friendly legislation adhere to the genetic view. The *Uniform Parentage Act* of 2000, a model law developed by the National Conference of Commissioners on Uniform State Laws, bars proceedings to adjudicate parentage if the child is more than 2 years old and there is a presumed father—a man who is married to the mother and who either cohabited or had sex with the mother around the time of the child's conception.[33] The *Act* also allows courts to deny requests for genetic testing at any point, depending on the effect that disestablishing paternity might have on the child and on various details about the relationship between the child and the man.[34] However, the traditional marital presumption is still the central mechanism by which denying genetic testing is justified.[35] An unmarried father, even if he accepts the child in his home, cares for her, supports her, and holds himself out as her natural father, is not a presumed father; he becomes the child's legal father only if he follows the *Act*'s procedures for acknowledging paternity or if he is declared the legal father through an adjudication of paternity.[36] The UPA 2000 thus implicitly holds that the man's relationship to the child is rooted both in a genetic connection to the child and in the man's relationship to the child's mother. As long as this view of fatherhood is dominant, it will be only natural for married men to suppose that, if they divorce their wives and can show that they are not genetically related to the child, then they ought to be able to opt out of fatherhood, and they ought to be able to use genetic testing as the lever that pops them free.

A less conceptual muddle arises in a second appropriate constraint on genetic paternity. In addition to constraining how people use genetic testing, we may constrain *how testing is provided* to them. One kind of appropriate constraint on the provision of genetic testing is obvious: the laboratories that offer genetic paternity tests may and probably should be regulated to ensure that testing is accurate and results dependable, even if the regulation means that the tests are more expensive or are offered by fewer laboratories. Given the level of certainty possible with genetic testing, results should almost never be overturned by further testing. To permit inaccurate testing is, of course, to compound the disruption that testing causes.

We can also reasonably limit access to testing by requiring that both legal parents (or the legal parent, if there is only one) grant consent to paternity testing. In most cases, the legal parents will be the rearing parents as well as the presumptive genetic parents. If the parental relationship is understood as a rearing relationship, then an existing rearing relationship should not be undermined secretly by someone else—whether a spouse, another family member, or a putative biological father outside the marriage.

There are some other, firmer ways of limiting access to genetic paternity testing. At the very end of the scale, we could consider prohibiting genetic testing altogether in certain categories of cases. Or we could make genetic testing permissible only when there is a court order for it. However, given the medical and sometimes social value of knowing who one's genetic relations are, this measure seems unwise. Further, whereas paternity testing that is intended to *reassess* parent-child relationships merits constraint, testing at birth, with the goal of initially sorting out parent-child relationships, presents a somewhat different set of concerns. The argument of this paper provides reasons for permitting such testing, even for encouraging it in those cases in which the parents themselves raise questions about genetic paternity: It would help children acquire the medical benefits of knowing their genetic ancestry. It could give biological parents an opportunity to "seize the moment" and establish a strong rearing relationship. Although it would disrupt some families that otherwise would never have discovered misattributed paternity, it might help avoid the disruption of later disputes. Given the rising importance of genetics in medicine, it may well be that an ever larger percentage of cases of misattributed paternity will be uncovered inadvertently anyway.

What argues most strongly against ever encouraging testing at birth is just that it would cultivate the very confusions that draw people to testing later on. If we could someday demystify the genetic relationship, recognize it for what it is, and understand both its appeal and its limits, we might even encourage testing at birth as a general social policy. But that day is not at hand.

NOTES

I am grateful to the participants of Genetic Ties and the Future of the Family, a research project run conjointly by The Hastings Center and the Institute for Bioethics, Health Policy and Law at the University of Louisville School of Medicine. Discussions held in the course of this project have influenced this paper in various ways. I am especially grateful to Mary Anderlik for detailed comments. An earlier version of this paper was presented at the 2002 annual meeting of the American Society of Bioethics and Humanities and at a colloquium at Oxford University on Genetic Technologies and the Family, and the paper has benefited from comments offered on each of those occasions. Funding for Genetic Ties and the Future of the Family is provided by the National Institutes of Health (grant #HG02485).

1. *Michael H. v. Gerald D.,* 491 U.S. 110, 24 (1989) (citing Nicholas H. *Adulturine Bastardy* 1836; 1:9–10).

2. Lewin T. In genetic testing for paternity, law often lags behind science. *New York Times* 2001 Mar 11:A1.

3. See note 2, Lewin 2001. See also: Stanley A. So, who's your daddy? in DNA tests, TV finds elixir to raise ratings. *New York Times* 2002 Mar 19:C1. The media coverage is also discussed at length in: Nelkin D. Paternity palaver in the media: selling identity tests. Unpublished manuscript, Genetic Ties and the Future of the Family, University of Louisville Center for Bioethics and The Hastings Center.

4. See, for example: http://www.dnacenter.com; http://www.genetree.com; and http://www. betagenetics.com.

5. Anderlik M. Disestablishment suits: daddy no more? Unpublished manuscript, Genetic Ties and the Future of the Family, University of Louisville Center for Bioethics and The Hastings Center. See: 1999 Ohio H.B. 242 (2000 Jul 27).

6. Egerton B. DNA tests don't let dads off the hook: man supports sons not biologically his. *Dallas Morning News* 1999 Oct 31:A1.

7. Grossberg M. How to give the present a past? family law in the United States 1950–2000. In: Katz SN, Eekelaar J, MacLean M, eds. *Cross Currents: Family Law*

and Policy in the United States and England. New York: Oxford University Press; 2000:3–29.

8. See note 7, Grossberg 2000:4–12.

9. See note 6, Egerton 1999.

10. See, for example: *Paternity of Cheryl,* 746 N.E. 2d. 488 (Mass. 2001); *NPA v. WBA,* 380 S.E. 2d. 178 (Va. Ct. App. 1989); *Dews v. Dews,* 632 A. 2d. 1160 (D.C. 1993).

11. Anderlik MR, Rothstein MA. DNA-based identity testing and the future of the family: a research agenda. *American Journal of Law and Medicine* 2002;28:215–32, at 218.

12. See, for example: *Michael H. v. Gerald D.,* 491 U.S. 110 (1989); *Rodney F. v. Karen M.,* 71 Cal. Rptr. 2d. 399,401 (Cal. Ct. App: 1998).

13. See, for example: *Cavanaugh v. deBaudiniere,* 493 N.W. 2d. 197, 197 (Neb. Ct. App. 1992); *Marriage of K.E. V,* 883 P. 2d. 1246 (Mont. 1994).

14. *William Carl Langston v. Alice L. Riffe; William Carl Langston v. Sharon Lock- lear; Danielle R. v. Tryone W.,* 359 Md. 396; 754 A. 2d. 389 (Md. 2000). Testing for such uses is also promoted; see http://www.dnagenetesting.com.

15. See: Grossberg M. *Governing the Hearth: Law and Family in Nineteenth- Century America.* Chapel Hill: University of North Carolina Press; 1985; Murray T. *The Worth of a Child.* Berkeley: University of California Press; 1996.

16. Pufendorf S. *On the Law of Nature and Nations.* Oldfather C, Oldfather W, trans. 1934:ch. 2, § 4.

17. Finkler K. *Experiencing the New Genetics: Family and Kinship on the Medical Frontier.* Philadelphia: University of Pennsylvania Press; 2000:127. Emphasis in the original.

18. Lifton BJ. *Journey of the Adopted Self: A Quest for Wholeness.* New York: Ba- sic Books; 1994, quoted in: Nelkin D. *The DNA Mystique: The Gene as a Cultural Icon.* San Francisco: Freeman; 1995:66.

19. For an extended discussion of this view and criticism of opposing views, see: Turkheimer E. Heritability and biological explanation. *Psychological Review* 1998;105:782–91.

20. Golombok S. *Parenting: What Really Counts?* London: Routledge; 2000:36.

21. See note 20, Golombok 2000:27–8.

22. See note 20, Golombok 2000:28–9.

23. Goldstein J, Freud A, Solnit A. *Beyond the Best Interests of the Child.* New York: Free Press; 1973:19.

24. See note 15, Murray 1996:57ff.

25. See note 15, Murray 1996:47ff. Murray draws from: Boswell J. *The Kindness of Strangers: The Abandonment of Children in Western Europe from Late Antiquity to the Renaissance.* New York: Pantheon; 1988.

26. Dolgin JL. Choice, tradition, and the new genetics: the fragmentation of the ideology of family. *Connecticut Law Review* 2000;32:529.

27. For an extended discussion of the relevance of these three interests to parent- hood, see: Blustein J. *Parents and Children.* New York: Oxford University Press; 1982:139ff.

28. The presumptive rules I identify are treated similarly by: Blustein J. Ethical issues in DNA-based paternity testing. Unpublished manuscript, Genetic Ties and the Future of the Family, University of Louisville Center for Bioethics and The Hastings Center.

29. See note 17, Finkler 2000:117ff.

30. Nelson JL. Genetic narratives: biology, stories, and the definition of the family. *Health Matrix* 1992;2:71–83.

31. A much more thorough discussion of the current law and policy options is found in: Rothstein M. Translating values and interests into the law of parentage determination. Unpublished manuscript, Genetic Ties and the Future of the Family, University of Louisville Center for Bioethics and The Hastings Center.

32. See note 5, Anderlik; 1999 Ohio H.B. 242 (2000 Jul 27).

33. National Conference of Commissioners on Uniform State Laws. *Uniform Parentage Act,* 2001 Jan 5 [UPA 2000]. The law also prohibits proceedings involving acknowledged or adjudicated fathers.

34. See note 33, UPA 2000, sect. 608.

35. Roberts P. *Biology and Beyond: The Case for Passage of the New* Uniform Parentage Act. Washington, D.C.: Center for Law and Social Policy; 2000:16ff.

36. Glennon T. Somebody's child: evaluating the erosion of the marital presumption of paternity, *West Virginia Law Review* 2000;102:547–605, at 569.

5

Ethics of Preimplantation Diagnosis for a Woman Destined to Develop Early-Onset Alzheimer Disease

Dena Towner and Roberta Springer Loewy

Since the birth of the first child conceived by in vitro fertilization more than 20 years ago, the applications of assisted reproduction have expanded rapidly. A single sperm can be injected directly into a single ovum to overcome severe male-factor infertility. An ovum from a young donor can be fertilized and implanted in a postmenopausal woman so she can carry the fetus. In vitro fertilization, in combination with DNA or karyotype analysis of a single cell from the developing embryo (preimplantation diagnosis), allows implantation of embryos that are free of genetic defects in couples without infertility. At every new step along the evolving pathway of assisted reproduction, ethical concerns have been raised. However, thus far the techniques have been permitted—in part because of a widespread but tacit assumption that promotion of reproductive freedom, or in today's language, reproductive autonomy—is an unqualified interest or good. But is reproductive freedom an unqualified good to society, the individuals undergoing the procedure, the medical profession, the businesses sponsoring such technologies, or the offspring thus produced?

Reproductive freedom is such a widely accepted norm in Western society that some even assume it to be an individual's absolute or inalienable right. But experience teaches that, like interests and goods, rights cannot be absolutely unqualified. First, there will always be disagreements about what these rights are, who grants them, and the source of their legitimacy.[1, 2] Even if everyone could agree on what they were, more than one right, interest, or good exists, guaranteeing that conflicting claims will occur and, if those conflicts are to be resolved, some negotiation will be necessary. Moreover, rights necessarily entail responsibilities, which also include respecting the rights of others and often qualify or even directly conflict with an individual's rights.[3-5] Thus, reproductive freedom, like other such ideas, is an ideal to work toward, and not a guaranteed right. In any issue that deals with a right, good, or responsibility, it is imperative that all of the values of all relevantly affected individuals be represented, articulated, and discussed. Otherwise, rights are merely being used like bludgeons or, as Churchill and Siman[6] call them, trump cards: whoever has the biggest stick or holds the trump card wins, and everyone else loses. Most such conflicts are not zero-sum games but are complicated balancing acts in which multiple competing claims need to be adjudicated so that risk to those most vulnerable or relevantly affected is minimized. But this only reflects that which is most characteristic of an ethical problem, ie, there is usually no good solution but a range of less than optimal choices, one of which must be chosen—since even not choosing is a choice by default.

The American Society for Reproductive Medicine Ethics Committee, anticipating such difficulties, has published reports to provide some guidance in some of these areas of concern. For instance, there is support for sex selection in preimplantation diagnosis when there is a medical indication such as Duchenne muscular dystrophy.[7] For some women, it is more acceptable to refrain from reimplanting affected embryos than to terminate a fetus diagnosed with a genetic disease. However, in current culture, sex selection is, under ordinary circumstances, to be discouraged.[7] Moreover, no physician is legally or ethically[8] bound to provide preimplantation sex selection services when requested by a fertile couple without a compelling medical (genetic) reason. In the interest of both the mother's and the child's optimal welfare, an upper age limit of 55 years for postmenopausal donor egg recipients has been proposed.[9] Thus, while compelling reasons must be provided to justify limiting reproductive freedom—such as the likelihood of serious harmful consequences or the existence of a stronger competing value—a wide range of possibilities for assisted reproduction remains.[7]

The study by Verlinsky and colleagues[10] in this issue of *The Journal of the American Medical Association* brings renewed concern to the issue of what does, or should, constitute ethically acceptable assisted reproduction. The authors have provided a woman with the opportunity to have a child free of an autosomal dominant form of Alzheimer disease (AD) that is encoded by the valine-to-leucine substitution at codon 717 mutation in the amyloid precursor protein gene. However, the mutation in her family confers onset of the disease during the fourth decade.[11] The woman was 30 years old when the procedure was performed, which means that she will likely manifest early symptoms of AD while this child is in the early, formative childhood years. In the report on this family describing the mutation,[11] the patient's sister was the relative in whom the mutation was first identified. One of the sister's first manifestations was difficulty in caring for her 2 children. By 5 years after the onset of symptoms, she was placed in an assisted-living facility. Much like her sister, the woman in the report by Verlinsky et al most likely will not be able to care for or even recognize her child in a few years. Analogously, women who are dying of cancer while their children are still young grieve greatly because they will not be able to see their children grow up, participate in their lives, or help protect them from harm. It is precisely these parental rights and responsibilities that are considered to be sufficiently compelling reasons for establishing an upper age limit for postmenopausal assisted reproduction.[12] Moreover, a child living under these circumstances would be burdened by the mother's progressive and eventually profound debilitation and eventual premature demise. Of note, if this same child were orphaned, current adoption regulations would prevent this same childless couple from adopting.

The family structure of this child also must be considered. If a healthy father (and possibly other extended family members) will be able to assume the role of primary care-giver for this child, these individuals presumably are also going to support and care for the woman as she progresses through the stages of AD. Thus, a single parent will be rearing this child in a few years. However, assisted reproduction involving a single parent is not uncommon. One example is a woman who uses her husband's frozen sperm after his death to have a child. This has been considered acceptable medical practice, provided the husband has given the semen sample knowing it may be used for this purpose in the future. Likewise, because single parenting is becoming more widespread, artificial insemination is performed for women who want a child but do not have a male partner.

However, there are not many situations in which assisted reproduction would be considered an option knowing the mother will not likely survive longer than 10 years. In most cases, the maternal medical condition limiting survival, such as pulmonary hypertension, advanced cancer, or end-stage renal disease, would preclude her from being a candidate for assisted reproduction. One could argue that a woman with human immunodeficiency virus (HIV) infection has a limited life expectancy. However, because survival time after a diagnosis of HIV infection is increasing with current treatment, it is uncertain whether a woman with HIV could reproduce and care for her child into adulthood. In fact, artificial insemination is used for couples in whom a woman is HIV-positive but her partner is not, to help prevent HIV transmission to the partner. Healthy HIV-positive women are not excluded from in vitro fertilization programs provided they understand the risk of HIV transmission to their infants, which has declined from 30% without any preventive measures to about 1% with use of anti-retroviral agents during pregnancy and cesarean delivery before onset of labor.[13] Alternatively, once a woman has a diagnosis of acquired immunodeficiency syndrome, assisted reproduction might not be offered due to the stress a pregnancy would place on her health in addition to the risks to the fetus of transmission.

Precisely because so many values and interests are in conflict in such cases, interest in the ethics of technology, including the interface between genetics and assisted reproduction, has increased considerably.[14-16] Although the report by Verlinsky et al raises several important ethical issues, this parent's ethical responsibility can be interpreted in at least 2 ways. One interpretation is that by resorting to selective preimplantation, the prospective parent was behaving in an ethically responsible manner by conceiving a child free of her own genetic predisposition for early-onset AD. While this is a laudable goal, it is not sufficient to satisfy this mother's ethical obligations because this interpretation is so narrow; it defines this woman's ethical responsibility solely in terms of disease prevention. Presumably, ethical responsibility is much broader. An alternative interpretation questions the purposive choice of bringing into the world a child for whom the mother will, with near certainty, be unable to provide care. The differences between these 2 interpretations of ethical responsibility are stark, but both rest on assumptions made about reproduction—is it a privilege or is it an unquestionable and inalienable right? Is it a mere want, a deeply held desire, or a need so profound and fundamental as to trump the rights or needs of others? While it may be extremely uncomfortable to question some of the most cherished and deepest-held assumptions, not doing so risks narrowing and distorting our understanding of a situation, the range of available alternatives, and their consequences, thereby reducing the ability to craft responsive and responsible decisions.

Ultimately, patients and physicians are faced with the "technology question": should a procedure be done simply because it can be done—and the subsidiary ones: who is to decide, and when? Because societal values include equity and respect for persons, the goal is to represent the interests of all relevantly affected. Certainly, this is not the responsibility of health care professionals alone. However, taking the responsibilities of beneficence and nonmaleficence seriously and moving beyond individualistic, subjective assumptions about such issues, physicians and other health care professionals must be willing to help patients and society understand, articulate, clarify, and discuss these different interpretations and their implications. By doing so, the burdens of those who will be most relevantly affected by these novel technologies—and who are often unable adequately to speak for themselves—will at least be minimized.

NOTES

1. Ross WD. *The Right and the Good.* Oxford, England: Oxford University Press; 1930.

2. Hart HL. Are there any natural rights? *Phil Rev.* 1955;64:175–191.

3. Dewey J. *Theory of the Moral Life.* New York, NY: Holt Reinhardt & Winston; 1960.

4. Rachels J. *The Elements of Moral Philosophy.* New York, NY: McGraw-Hill; 1993.

5. Loewy EH. Communities, obligations and healthcare. *Soc Sci Med.* 1987;25: 783–791.

6. Churchill LA, Simán JJ. Abortion and the rhetoric of individual rights. *Hastings Cent Rep.* 1982;12:9–12.

7. The Ethics Committee of the American Society of Reproductive Medicine. Sex selection and preimplantation diagnosis. *Fertil Steril.* 1999;72:595–598.

8. The Ethics Committee of the American Society of Reproductive Medicine. Preconception gender selection for nonmedical reasons. *Fertil Steril.* 2001; 75:861–864.

9. Fasouliotis SJ, Schenker JG. Ethics and assisted reproduction. *Eur J Obstet Gynecology Reprod Biol.* 2000;90:171–180.

10. Verlinsky Y, Rechitsky S, Verlinsky O, Masciangelo C, Lederer K, Kuliev A. Preimplantatlon diagnosis for early-onset Alzheimer disease caused by V717l mutation. *JAMA.* 2002;287:1018–1021.

11. Murrell JR, Hake AM, Quaid KA, Farlow MR, Ghetti B. Early onset Alzheimer disease caused by a new mutation (V717l) in the amyloid precursor protein gene. *Arch Neurol.* 2000;57:885–887.

12. Schenker JG. Ovum donation: ethical and legal aspects. *J Assist Reprod Genet.* 1992;8:411–418.

13. ACOG committee opinion scheduled Cesarean delivery and the prevention of vertical transmission of HIV Infection: number 234, May 2000 (replaces number 219, August 1999). *Int J Gynaecol Obstet.* 2001;73:279–281.

14. Roberts MA. *Child Versus Childmaker: Future Persons and Present Duties in Ethics and the Law.* Lanham, Md: Rowman & Littlefield; 1998.

15. Cooper SL, Glazer ES. *Choosing Assisted Reproduction: Social, Emotional and Ethical Considerations.* Indianapolis, Ind: Perspectives Press; 1998.

16. Bayertz K, ed. *The Concept* of *Moral Consensus: The Case* of *Technological Interventions in Human Reproduction.* Boston, Mass: Kluwer Academic Publishers; 1994.

6

Procreation for Donation: The Moral and Political Permissibility of "Having a Child to Save a Child"

Mark P. Aulisio, Thomas May, and Geoffrey D. Block

INTRODUCTION

The crisis in donor organ and tissue supply is one of the most difficult challenges for transplant today. New policy initiatives, such as the driver's license option and *required request,* have been implemented in many states, with other initiatives, such as *mandated choice* and *presumed consent,* proposed in the hopes of ameliorating this crisis. At the same time, traditional acquisition of organs from human cadavers has been augmented by living human donors, and nonheartbeating human donors, as well as experimental animal and artificial sources. Despite these efforts, the crisis persists and is perhaps most tragic when it threatens the lives of children, driving parents to sometimes desperate measures. Herein, we address one very controversial step some parents have taken to obtain matching tissue or organs for their needy children—that is, having a child, in part, for the purpose of organ or tissue procurement.[1]

One might address this issue at either of two levels, that of individual choice or social policy. At the level of individual choice, one could ask, for example, whether or not parents faced with such a decision should choose to bring another child into the world partly for this purpose. This level of discussion requires, among other things, surfacing considerations that parents should take into account in decisionmaking as well as an evaluation of those considerations from some particular substantive moral viewpoint. The level of social policy, however, is quite different. Here the question is not whether any particular parents faced with such a difficult decision should choose one way or another, but rather whether parents in these circumstances should be *permitted* to choose at all. It is important to underscore at the outset that our focus is on the latter—that is, the moral and political *permissibility* of "having a child to save a child" in our *societal* context—and *not* on whether the practice should be morally forbidden, permitted, or praised from some particular substantive moral perspective that different individuals in our society might adopt (e.g., the perspective of any particular religious tradition). Thus, one could support the permissibility of having a child to save a child given our societal context and still consider it to be immoral in light of one's own values and particular moral views. In short, the question we focus on is: Should parents in our society be morally and politically *permitted* to have a child so that it can be the source of tissue or organs needed to save the life of one of their other already existing children?

THE CASE OF THE AYALA FAMILY

Cases of having a child to save a child are a relatively recent phenomenon, emerging with the advent of improved techniques in transplant and the emphasis on HLA matching for optimal outcomes.[2] The most common reported cases involve the need for matching bone marrow, but there is no *medical* reason that precludes having children as a source of organs, such as kidneys, lung, intestines, or liver. When we speak of having a child to save a child, we have in mind the following type of case, as described by Gail McBride in her 1990 *British Medical Journal* article:

> Forty-five-year-old Abe Ayala, who had had a vasectomy 17 years ago, had it reversed so that he and his wife, Mary, age 43, could conceive a child. A major reason for the decision was the one-in-four chance that the infant could serve as a bone marrow donor for the Ayalas' 18-year-old daughter, Anissa, who has chronic myelogenous leukemia. . . .
>
> The news that the family had taken the step of conceiving another child was greeted by a storm of criticism from ethicists in the United States, who said that the infant was created not as an end in itself but for a utilitarian purpose and that, moreover, the infant could not give informed consent for any bone-marrow transplantation procedure.[3]

McBride does not quote any particular ethicist who espoused this view at the time and there are few published articles on either side of the subject. Nonetheless, the case raises a number of interesting and important medical, moral, and political issues, including questions about

- acceptable motivations or reasons for having children,
- the use of children as a source of organ and tissue,
- possible risks and benefits to children who donate,
- the likelihood of benefit to sibling recipients, and
- limits of parental rights to consent to donation on behalf of children.

MINORS AND LIVING-RELATED DONATION

The controversial having a child to save a child cases bear important similarities to cases of minors and living-related donation. The most important similarity, of course, exists at the point of donation. Here having a child to save a child cases *just are* "minors and living-related donation" cases. Given that there is a great deal of consensus about the latter, it will be helpful to see how they have been dealt with and whether this sheds any light on the controversial having a child to save a child cases with which we are concerned.

U.S. courts have dealt with a number of minors and living-related donation cases. We will consider three: *Hart v. Brown* (1972), *Curran v. Bosze* (1990), and *Strunk v. Strunk* (1969).

Hart v. Brown (1972)

The case of *Hart v. Brown* involved 7-year-old identical twins, Kathleen and Margaret Hart. Kathleen developed hemolytic-uremic syndrome, which led to malignant hypertension and kidney failure. The twins' parents, Peter and Eleanor Hart, wanted to give permission on behalf of Margaret to donate a kidney to Kathleen. The Yale-New Haven Hospital and its surgeons, however, were unwilling to go forward with the transplant without a ruling by the court on the rights of the parents or guardians to give consent on behalf of Margaret Hart.[4]

On purely medical grounds, transplant from Margaret was uncontroversially in Kathleen's interest. Being identical twins, the donated organ would have a perfect HLA match with the recipient. Further, the shortened time between recovery and reestablishment of blood flow would have positive benefits on organ function. Although the latter benefit could be expected with donation by one of the parents, HLA matching would only be 50% (or a three-antigen match). The National Kidney Transplant Database System and UNOS data indicate improvement in 5-year survival of 3–7% between a living-related donation (parent in this case) and a perfect HLA match (twin sister) even with today's vastly superior antirejection drugs.[5] Historical data and experience support the concept that identical twins have additional matching within the immune system such that antirejection-drug requirements are often significantly less than with unrelated, perfect HLA match, donor-recipient pairs. On the other hand, it is equally clear that, on strictly medical grounds, the removal of a kidney from Margaret was not in her interest. Whereas the immediate risks of living-related kidney donation are small (<1% mortality or significant morbidity), there is nonetheless a real and present risk of significant medical consequence. Further, there are cosmetic consequences and potential impact on access or acceptability to childhood and adolescent social, school, and athletic opportunities. Additionally, there are a number of medical illnesses (such as diabetes and hypertension) that might one day occur and adversely affect renal function that could be more clinically significant with one kidney versus two kidneys.[6]

The Court, however, did not decide the case on the basis of the medically optimal course for the sickest child. Nor did it decide the case on the basis of the competing interests of Kathleen and Margaret, allowing Kathleen's survival interests to trump the risk to Margaret's interests posed by donation or, as some have suggested, on the basis of the *moral* significance of familial relationships.[7] Rather, the Court held that Peter and Eleanor Hart had the right to consent to donation on behalf of Margaret because it was in *Margaret's* "best interest" to donate.[8]

The Court's decision may seem, at first glance, quite counterintuitive to some.[9] How could it be in the best interest of a perfectly healthy 7-year-old girl to give up one of her kidneys? The Superior Court of Connecticut appealed to two legal doctrines, the doctrines of "substituted judgment" and "grave emotional impact," along with the testimony of clergy, psychiatrists, and other experts to arrive at its decision.[10] We will discuss the doctrine of substituted judgment in more detail later. The key point now, for our purposes, is that the Court deemed that Margaret would be so emotionally devastated by the loss of Kathleen ("grave emotional impact") that it was, on balance, *in her interest* to donate.[11] The devastation, the Court reasoned, would be manifest directly in the loss of her twin sister, to whom she wished to donate, and indirectly through the effect of Kathleen's death on the family.[12]

We will refer to this rationale as the best interest standard. Later we will consider the *propriety* of the best interest standard for cases of minors and living-related donation. For now, however, it is sufficient to recognize that this was the basis of the Court's decision.

Curran v. Bosze (1990)

The best interest standard was also invoked in a more recent case before the Supreme Court of Illinois but with a very different outcome. In 1988, 8-year-old Jean Pierre Bosze was diagnosed with acute undifferentiated leukemia, which went into remission and then recurred after 2 years.[13] By 1990, Jean Pierre was in need of a bone marrow transplant. His father, Tamas Bosze, mother, and 23-year-old stepbrother were not HLA identical. Mr. Bosze, several years prior, had fathered twins from an affair with Mrs. Nancy Curran.

Mrs. Curran refused to allow the twins, by then 3 years old, to be tested for compatibility. Mr. Bosze filed suit to have the twins submit for bone marrow harvesting against the wishes of their mother, Mrs. Curran.[14]

Here we again have a clash of medical interests. On the one hand, it was clearly in Jean Pierre's interest to have the twins tested for compatibility. On the other, the twins would have no medical benefit and some risk of harm by submitting to testing.[15]

Interestingly, the Supreme Court of Illinois *refused* to compel the twins to submit to bone marrow harvesting. The Court held that the doctrine of substituted judgment was inapplicable and that the best interest of the twins was not served by compelling them to submit to the procedure.[16] The Court specifically cited the presence of "an existing close relationship between the donor and the recipient" as one of three determinative factors in judging what is in the best interest of a child who might donate to a sibling (the other two factors being genuine informed consent by the parent or guardian, and the availability of emotional support to the child from caregivers).[17]

Strunk v. Strunk (1969)

The best interest standard played an equally critical role in the 1969 Kentucky case of *Strunk v. Strunk.* Twenty-eight-year-old Tommy Strunk suffered from chronic glomerulus nephritis and was in need of kidney transplant. Tommy had an institutionalized 27-year-old half-brother, Jerry, with an IQ of 35. Mrs. Strunk, their mother, wanted to consent on Jerry's behalf to donate a kidney to Tommy, in order to save his life. Mrs. Strunk petitioned the court for the right to consent on Jerry's behalf.[18]

This case bears some important similarities and differences to each of the preceding cases. Like *Hart*, the case involves a potential kidney donation. Unlike *Hart*, but like *Bosze*, Jerry and Tommy Strunk are only half-siblings. Strunk differs from *both Hart* and *Bosze* in that neither Jerry nor Tommy is a minor. However, Jerry is legally incompetent and therefore has the standing of a minor for the purposes of law and morals.

This latter similarity, coupled with presence of an "existing close relationship" between Jerry and Tommy proved to be decisive in determining the outcome of the case. Ultimately, the Kentucky Court of Appeals upheld the decisions of the lower courts to permit Mrs. Strunk to consent to donation on Jerry's behalf.[19] The courts determined on the basis of testimony from a variety of sources, including psychiatrists and an amicus curiae filed by the Kentucky Department of Mental Health, that the death of Tommy under these circumstances would have "an extremely traumatic effect upon"

Jerry.[20] Indeed, the amicus curiae stated that Jerry " . . . identifies with his brother Tom; Tom is his model, his tie with his family . . . Tom's life is vital to Jerry's improvement. . . ."[21] Thus, as in the Hart case, the fact that the courts deemed donation to be, on balance, in Jerry's best interest provided the justificatory grounds for the decision.

MORAL AND POLITICAL PROPRIETY
OF THE BEST INTEREST STANDARD

Some have challenged the moral and political propriety of the best interest standard in cases of minors and living-related donation.[22] We suggest, however, that the best interest standard is morally and politically appropriate given the societal context in which these types of decisions must be made. Furthermore, we suggest, in the next section, that the best interest standard is morally and politically appropriate for decisionmaking in the controversial having a child to save a child cases as well. Is the best interest standard morally and politically appropriate for cases of minors and living-related donation?

The Strunk case is instructive for our purposes not only because it involves living-related donation but also because it involves a legally *incompetent* adult, Jerry Strunk. *Competent* adults, in our society, have a right, stemming from the societal value of autonomy, to live their lives according to their values. These values oftentimes have implications for medical decisionmaking. For example, among the range of medically acceptable options, judgments of "best treatment" are value laden, often contingent on the particular values of the patient. Similarly, the right of a competent patient to refuse treatment is rooted in a recognition of the value-laden nature of "quality of life" and an individual's right to live her life as she sees fit.[23] The importance of respecting patient values in medical decisionmaking is why informed consent matters and why advance directives are valuable tools for decisionmaking at the end of life. Within certain limits, the decision to donate tissue or organs is, therefore, up to competent adults as well. Indeed, the gold standard here, as in other areas of medical decisionmaking, ought to be genuine informed consent.[24]

When adults are not competent, however, things are less clear. To the extent possible the values and preferences when competent of a once competent, but now incompetent, adult should be respected. This, as mentioned above, grounds the utility and justifiability of advance directives, and it should also inform a proper understanding of the role of surrogates.[25] As the Court noted in the Bosze case, the doctrine of substituted judgment properly understood *just is* the attempt to assess the implications of a once competent, now incompetent, individual's values and preferences (when they were competent) for decisionmaking. *There is a profound sense in which best interest for competent adults (and for once competent, but now incompetent, adults) must be defined where possible by their values and preferences (or the values and preferences they expressed when competent).* Thus, substituted judgment is applicable and appropriate in decisions to donate organs to the extent that the values (when competent) of a once competent adult are known and have implications for donation.

The category of cases including (1) once competent adults, whose values and preferences were not known, (2) now incompetent adults, who were never competent, and (3) children, poses its own set of difficulties. Jerry Strunk, the Bosze twins, and the Hart twins, of course, fall into this category. Precisely because there is no way to define "best interest" for these persons in light of their autonomous values and preferences, it must be defined in a quasi-objective manner. The State, therefore, adopts a paternalistic role with respect to their well-being and so is primarily concerned with their best interest in a more objective sense. This State role stems from its interest in the prevention of harm to innocent persons (the most basic limit on the liberties of autonomous individuals).[26] Indeed, even though parents' or guardians' judgments of best interest often drive decisionmaking, the scope of these judgments is itself limited by a quasi-objective notion of best interest. This is what grounds, for example, intervention in cases of child abuse or neglect and limits the role of parental beliefs (religious or other) in refusals of treatment for their children. In the context of organ donation, one can imagine cases involving parents or guardians who might want to sell one of their children's, or never competent adult's, organs. This would be forbidden under any plausible interpretation of best interest. (In the United States and much of the world, with the exception of sperm and eggs, the sale of human tissue is illegal.)

Given this societal context, which includes the autonomy rights of individuals and the State interest in preventing harm to innocents, it is our view that the best interest standard is morally and politically appropriate in cases of minors (or never competent adults) and living-related donation. Interestingly, in the Hart and Strunk cases, the Courts interpreted best interest in a manner akin to substituted judgment by taking into account the existing other-regarding values and preferences of Margaret Hart and Jerry Strunk, despite the fact that they were not (and never had been) legally competent. The Courts' interpretation, therefore, allowed for the values and preferences of Margaret and Jerry to at least partly define best interest.

"HAVING A CHILD TO SAVE A CHILD" CASES REVISITED

We now have a background against which to judge having a child to save a child cases. Because these cases involve minors and living-related donors, we have at least a prima facie reason to handle them in a similar way, provided there is no salient difference that would justify handling them differently. There is, however, one salient and obvious difference: In having a child to save a child cases, the prospective donor child is brought into existence, at least in part, for the *purpose* of organ or tissue procurement; or, to put it in a more rhetorically compelling way, the child is being brought into the world, at least in part, so that it can be *used,* in effect as a *commodity* (i.e., as a source of organ or tissue). Indeed, McBride describes the argument that "the infant was created not as an end in itself, but for a utilitarian purpose" as the primary concern of ethicists who criticized having a child to save a child at the time the Ayalas' case became public.[27] Is this difference between minors and living-related donation cases and having a child to save a child cases sufficient to distinguish the two types of cases and justify proscription of the latter for the purposes of social policy?

This difference, of course, essentially involves a challenge to the reasons parents have for bringing a new child into the world in having a child to save a child cases. In discussing reasons for having children, we need to distinguish between morally acceptable or good reasons from any particular individual or community's substantive moral viewpoint, and reasons that fall within the realm of the societally permissible. As we stressed at the outset, the latter is our primary concern. Though few would argue that conceiving a child to provide organs or tissue for transplantation is an optimal reason for having a child, the important question for our purposes here is whether there is sufficient justification for prohibiting this practice in our societal context—a context that, for better or worse, sets very few limits on individuals' reproductive rights.

People have children for many different reasons (to save a marriage or relationship, to have something to hold, to placate potential grandparents, etc.), and often for no particular reason at all (some recent studies suggest that almost half of all pregnancies in the United States are unplanned, with just under half of these pregnancies being brought to term).[28] Although having a child to provide matching organs or tissue may be less than optimal, having children for certain other reasons, or no reason at all, is also less than optimal. Indeed, a strong argument could be made that having a child to save a child is a better reason to have a child than any of the above, yet we do not, as a society, prohibit procreation for these other reasons. Given this societal context, it is very difficult to find any justificatory ground on the basis of reasons or motivations for prohibiting persons like the Ayalas from bringing a child into the world in the hopes of saving one of their other children, without serious ramifications for reproductive rights across the board.

Though to many it is far from obvious that there is anything objectionable about the intent, as characterized, of parents in having a child to save a child cases like that of the Ayala family, it may be that their intention is being mis-described. "Intention" in both law and morals is a notoriously slippery concept that must be carefully parsed.[29] The line between intention and foresight is a fine one, and in practical contexts it can be very difficult to draw. One could plausibly describe the Ayalas' intention, for example, as "promoting life by bringing another child into the world, while foreseeing that this will also have the effect of giving their existing child a better chance at obtaining matching tissue." One could invoke the principle of double effect, often used to justify palliative care at the end of life and recognized in law, to justify having a child to save a child. On this description, the Ayalas would have a morally acceptable end of affirming the good of procreation and an acceptable means of loving sexual relations. Obtaining matching tissue could be described as a foreseen and desired, but unintended, side effect. Under this plausible redescription of the Ayalas' intentions, their decision would satisfy the intention condition of double effect and, quite uncontroversially, the proportionate reason condition.[30] Thus, it is far from clear that there is anything objectionable (i.e., that amounts to "using" or "commodifying") about the intentions of parents in having a child to save a child cases in the first place, quite apart from considerations of the implications of proscription for reproductive rights in general.

It might be argued, however, that a child brought into the world, in part, to be a tissue or organ donor will not be loved for his or her own sake and is thus likely to be abused or neglected. This strikes us as, at best, a dubious empirical claim. Indeed, the numbers of births resulting from unplanned pregnancies or pregnancies for less than optimal reasons constitutes a significant percentage of overall births in the United States, yet we know of no evidence suggesting that a disproportionate number of these children are neglected or abused (and there surely is no reliable evidence of this with respect to the tiny class of cases we are concerned with here). Even if there were evidence to suggest that children conceived for less than optimal reasons or no reason at all were more likely to be abused or neglected, it is not clear how this should affect social policy regarding reproductive rights anyway.[31] Furthermore, it seems reasonable to assume that parents who care enough about an existing child to bring another child into existence partly in the hopes of saving the existing child are likely to love and care for their new child as well (or at least as likely to do so as parents who have children for other less than optimal reasons or no reason at all).

Interestingly, once the issue of permissible reasons for having a child is dismissed, we are, in our view, left with no remaining salient difference between the controversial having a child to save a child cases and cases of minors and living-related donation about which there is a great deal of consensus. Indeed, once a child has been brought into the world, having a child to save a child cases de facto become minors and living-related donation cases. Earlier we argued in favor of a "best interest of the potential donor" standard as morally and politically appropriate for cases of minors and living-related donation. If we are correct, then having a child to save a child cases, at the point of possible donation, should be subject to the same standard to determine permissibility. Is donation in these cases likely to be in the best interest of the prospective donor child?

The answer to this question is far from obvious. In the Hart and Strunk cases, for example, the courts appealed to the "presence of an existing close relationship" between the siblings as a key component in determining best interest. In cases of bone-marrow harvesting from infants, like that of the Ayala family, an existing close relationship would not be present in any robust sense. Given the relatively low risks of bone-marrow harvesting, however, and the arguably significant benefits of growing up in a closely knit family with a healthy brother or sister, it seems plausible to construe donation as, ceteris paribus, in a child's best interest. Indeed, the long-term effects of losing a sibling both directly on the new prospective donor child (who would grow up without that sibling) and on the family in which the prospective donor child will be raised are likely to be very significant.[32]

As the risks to the prospective donor child escalate, however, as they would for organ donation, the absence of an existing close relationship becomes even more problematic.[33] Some might be tempted to argue that despite an escalating risk to the prospective donor child and the absence of an existing close relationship, donation in having a child to save a child cases would still be justified under a best interest of the donor standard because the donor child would not even exist *but for* the need for donation. Because it is better to be without a kidney than not to be at all (to answer a variation on Shakespeare's famous query), so the argument might go, donation would remain on balance clearly in the prospective donor child's best interest. Will it almost always, then, be in the best interest of the prospective donor child to donate in having a child to save a child cases?

This argument exhibits a dangerous misunderstanding of the best interest of the donor standard. As discussed earlier, the best interest standard exists to protect innocents from harm. Thus, once a child actually exists, a determination of whether donation is in the child's best interest must be made. If donation is not in a child's best interest at that time, then it may not go forward. In having a child to save a child cases this means determining whether or not donation will be in the prospective donor child's best interest once the prospective donor child actually exists and donation is on the horizon. To use the claim that "it is better to be than not to be" to establish best interest at the point of donation would open the door to visit almost any type of harm on children, provided visiting that harm was made a condition of bringing them into the world in the first place. For example, on this twisted version of the best interest standard, parents could have a child for the purpose of donating (or perhaps even selling) both of the child's kidneys to science, consigning the child to a dialysis-dependent life, and still argue that, on balance, donation was in the child's best interest (for a life on dialysis is better than no life at all . . .).

This being said, we do think that best interest in the Ayala case and others like it can be plausibly construed as in favor of donation, much as it was in the Hart and Strunk cases. However, for our purposes here, there are two key points. First, as a matter of social policy there is no justificatory ground for prohibiting parents from having a child to save a child. Second, whether or not donation may go forward in any particular having a child to save a child case should depend on satisfying a best interest of the donor standard at the time of donation just as it does in cases of minors and living-related donation. This, of course, will have to be considered on a case-by-case basis with attention to the unique circumstances of each particular case.[34]

CONCLUSIONS AND CAUTIONS

Though we support the moral and political *permissibility* of having a child to save a child and moving forward with donation, provided it is in the best interest of the prospective donor child at the time of donation, we want to conclude with a word of caution. Some, of course, will be concerned about having a child to save a child cases, the dangerous precedent these cases might set, and their potential to open Pandora's box. There are indeed some variations on having a child to save a child cases that raise disturbing questions. For example, suppose that parents in a having a child to save a child case decide that they will abort or give the child up for adoption if it is not a match. Should this affect the moral and political permissibility of having a child to save a child? One can also easily imagine cases of women getting pregnant not with the intention of bringing the pregnancy to term, but rather to sell embryonic cells, fetal tissue, or organs ("having a pregnancy to sell organs and tissue"). This raises a whole constellation of problematic questions that would have to be examined as society moves toward permission or proscription.

Similarly, technological developments might allow having a child to save a child to give way to "cloning a child to save a child." Current technology with in vitro fertilization methods already provides a legal means for effectively creating an HLA-matched embryo, fetus, or child. This too raises a complex constellation of issues that would have to be addressed to provide the justificatory grounds for permissibility or proscription.

Although these variations are disturbing and raise questions warranting serious public discussion, we believe it is imperative to treat like cases alike, and we do not think having a child to save a child in itself thrusts one down the slippery slope. Herein, we have not addressed these important questions raised by variations on having a child to save a child cases. Our concern has been more limited, focusing specifically on the moral and political permissibility of having a child to save a child cases themselves. We have argued that the best interest standard is appropriate for having a child to save a child cases, such as that of the Ayala family, just as it is for cases of minors and living-related donation. We have done this by defending the best interest standard as morally and politically appropriate for resolving cases of minors and living-related donation, and showing that there are no salient features of having a child to save a child cases that would justify treating them differently. This does not imply that there are no salient features of *variations* on having a child to save a child cases that might warrant treating them differently. That discussion, however, must wait for another day.

NOTES

1. Degrees (and even types) of risks will vary greatly for the procurement of different organs and tissue and among prospective donors. This fact does not affect our argument regarding the appropriate moral and political standard that should drive decisionmaking but affects, rather, whether the standard is satisfied in any particular case. Thus, we do not distinguish between these for the purposes of this article.

2. McBride G. Keeping bone marrow donation in the family. *British Medical Journal* 1990;300:1224–5.

3. See note 2, McBride 1990. It is important to emphasize that we are addressing having a child to save a child cases like that of the Ayala family, and not somewhat related, but importantly different, classes of cases like "having a pregnancy to sell organs or tissue" or "cloning a child to save a child" and so on. We take the case of the Ayala family as paradigmatic of the class of cases with which we are concerned herein.

4. *Hart v. Brown*, 289 A. 2d 386 (1972).

5. Thus, in 1972, of course, the difference in survival rates for these two groups was much more pronounced. For 5-year graft survival rates of cadaveric, living-related, and living-unrelated donor kidney transplants, see the UNOS 1998 Annual Report, Tables 28 and 30, available at: http://www.unos.org/data/anrpt98/ar98.

6. For an interesting recent study of U.S. kidney-transplant center policies toward children and living kidney donation, see: Spital A. Should children ever donate kidneys? views of U.S. transplant centers. *Transplantation* 1997;64:232–6.

7. See, for example: Dwyer J, Vig E. Rethinking transplantation between siblings. *Hastings Center Report* 1995;25:7–12. Dwyer and Vig suggest that the Court's discussion of the "best interest" of Margaret as a justification for allowing donation is merely a "rationalization" for a decision reached on other grounds (p. 9). Indeed, they state that "The obvious and direct question is whether the parents are justified in placing the healthy child at risk in order to meet the needs of the ill child" (p. 10). They then argue that it is moral significance of the "family relationship that justifies imposing some risks" on one child, "provided the expected benefits" to the other are substantial (p. 11). A similar position is taken in: Klepper H. Incompetent organ donors. *Journal of Social Philosophy* 1994;25:241–55. For efforts to emphasize the moral significance of the family relationship in other contexts, see: Hardwig J. What about the family? *Hastings Center Report* 1990;20:5–10; and Nelson JL. Taking families seriously. *Hastings Center Report* 1992;22:6–12.

8. For the full discussion by the Court, see note 2, *Hart v. Brown* 1972:386.

9. See, for example: Delany L. Protecting children from forced altruism: the legal approach. *British Medical Journal* 1996;312:240. Delany agrees that best interest of the child donor is the appropriate legal standard but expresses skepticism that, even in bone-marrow donation to a sibling, donation is actually in the child's best interest.

10. The Court appealed to the doctrine of "substituted judgment" as grounding its power to act in cases involving legal incompetents and, by extension, to minors (see note 2, *Hart v. Brown* 1972:289). More properly understood, "substituted judgment" has come to have a related, but different, meaning that we discuss later (see the section entitled "Moral and Political Propriety of the Best Interest Standard").

11. For example, the Court invoked the testimony of a psychiatrist in its rationale, stating that "the donor has a strong identification with her twin sister" and he (the psychiatrist) testified that if the expected successful results are achieved they would be of immense benefit to the donor in that the donor would be better off in a family that was happy than in a family that was distressed and in that it would be a very great loss to the donor if the potential recipient were to die from her illness *(Hart v. Brown, 289 A. 2d 389, 1972)*. For direct appeals by the Court to the doctrine of "grave emotional impact," see note 2, *Hart v. Brown* 1972:390–1.

12. See note 2, *Hart v. Brown* 1972.

13. See *Curran v. Bosze*, 566 N.E. 2d 1319 (Ill. 1990), for the Court's holding and rationale. This case received a great deal of media attention at the time (for example, see *Time Magazine,* 10 Sept 90, Vol. 136, Issue 11, p. 70).

14. See note 13, *Curran v. Bosze* 1990.

15. For a discussion of North American pediatric transplant centers views of minors and bone-marrow donation, see: Chan KW, Gajewski JL, Supkis D, Pentz R, Champlin R, Bleyer WA. Use of minors as bone marrow donors: current attitude and management. *The Journal of Pediatrics* 1996;128:644–8.

16. Indeed, the Court held that "(1) the doctrine of substituted judgment was inapplicable, and (2) it was not in the best interests of the twins to submit to the bone marrow harvesting procedure." See note 13, *Curran v. Bosze* 1990:1319. The Court rightly noted that the doctrine of substituted judgment concerns attempting to respect the values and preferences of a once competent, now incompetent, person in decisionmaking, thus implying that it had no implications for a case like *Hart v. Brown* in which it was, as we have seen, invoked. We discuss "substituted judgment" in more detail in the section entitled "Moral and Political Propriety of the Best Interest Standard."

17. See note 13, *Curran v. Bosze* 1990:1343.

18. *Strunk v. Strunk,* Ky., 445 S.W. 2d 145 (1969).

19. See note 18, *Strunk v. Strunk* 1969:145.

20. See note 18, *Strunk v. Strunk* 1969:146.

21. See note 18, *Strunk v. Strunk* 1969:146.

22. As we have seen, the most powerful challenges contend that the "best interest" standard is excessively individualistic and ignores important moral relationships between family members (see note 7, Hardwig 1990; Nelson 1992; and Dwyer, Vig 1995). These relationships are said to establish moral obligations on the part of some persons to donate, which obligations the courts should take into account in adjudication (see note 7, Dwyer, Vig 1995). Though we are sympathetic to the views expressed, the problem with this line of argument is that it must presume a particular substantive view of the good life that should be taken into account by the courts. The societal political framework within which these decisions must be made, however, does not (indeed may not) presume such a privileged view of the good, no matter how widely accepted or plausible it may be.

23. See Cardoza's famous decision in *Schloendorf v. Society of New York Hospital* (1914), a landmark case for the right to refuse treatment and the importance of informed consent in medicine.

24. For outstanding comprehensive discussions of the importance of informed consent in medical decisionmaking, see: Appelbaum PS, Lidz CW, Meisel A. *Informed Consent: Legal Theory and Clinical Practice,* London/New York: Oxford University Press, 1887 [*sic*]; and Faden RR, Beauchamp TL. *History and Theory of Informed Consent,* London/New York: Oxford University Press, 1996.

25. See Buchanan AE, Brock DW. *Deciding for Others: The Ethics of Surrogate Decision-Making,* London/New York: Cambridge University Press, 1989.

26. See Joel Feinberg's outstanding discussion of the harm principle and liberty in: *Harm to Others: The Moral Limits of the Criminal Law,* London/New York: Oxford University Press, 1986.

27. See note 4, McBride 1990:1224–5.

28. See Henshaw SK. Unintended pregnancy in the United States. *Family Planning Perspectives* 1998;30:24–9, 46.

29. See Aulisio MP. In defense of the intention/foresight distinction. *American Philosophical Quarterly* 1995;32:341–54; and Aulisio MP. On the importance of the intention/foresight distinction. *American Catholic Philosophical Quarterly* 1996;70:189–205.

30. For a general and concise discussion of the principle of double effect and its applications in healthcare, see: Beauchamp TL, Childress JF. *Principles of Biomedical Ethics,* 3rd ed. London/New York: Oxford University Press, 1989:127–34.

31. Surely, for example, a societal prohibition on unintended pregnancies being brought to term would be unjustified, even if data conclusively showed that children resulting from unintended pregnancies were more likely to be abused or neglected.

32. It should be noted, however, that even for this category of cases determinations of best interest of the prospective donor child must be made on a case-by-case basis because there could be complicating factors that make, on balance, even the harvesting of bone marrow not in the prospective donor child's best interest.

33. Unless, of course, an existing close relationship develops over a number of years, as it may in a having a child to save a child kidney-donation case.

34. It should be noted, of course, that as outcomes for cadaveric, living-unrelated, and various types of living-related (parent vs. sibling, etc.) organ donation become more even (as they have, for example, for kidney transplant), it is harder to justify donation from a minor in light of these alternatives. This, obviously, does not uniquely apply to having a child to save a child cases but rather it applies to all cases of minors and organ donation.

7

Population Screening in the Age of Genomic Medicine

Muin J. Khoury, Linda L. McCabe,
and Edward R. B. McCabe

Physicians in the era of genomic medicine will have the opportunity to move from intense, crisis-driven intervention to predictive medicine. Over the next decade or two, it seems likely that we will screen entire populations or specific subgroups for genetic information in order to target interventions to individual patients that will improve their health and prevent disease. Until now, population screening involving genetics has focused on the identification of persons with certain mendelian disorders before the appearance of symptoms and thus on the prevention of illness[1] (e.g., screening of newborns for phenylketonuria), the testing of selected populations for carrier status, and the use of prenatal diagnosis to reduce the frequency of disease in subsequent generations (e.g., screening to identify carriers of Tay-Sachs disease among Ashkenazi Jews). But in the future, genetic information will increasingly be used in population screening to determine individual susceptibility to common disorders such as heart disease, diabetes, and cancer. Such screening will identify groups at risk so that primary-prevention efforts (e.g., diet and exercise) or secondary-prevention efforts (early detection or pharmacologic intervention) can be limited. Such information could lead to the modification of screening recommendations, which are currently based on population averages (e.g., screening of people over 50 years of age for the early detection of colorectal cancer).[2]

In this review, we describe current and evolving principles of population screening in genetics. We also provide examples of issues related to screening in the era of genomic medicine.

PRINCIPLES OF POPULATION SCREENING

The principles of population screening developed in 1968 by Wilson and Jungner[3] form a basis for applying genetics in population screening. These principles emphasize the importance of a given condition to public health, the availability of an effective screening test, the availability of treatment to prevent disease during a latent period, and cost considerations. Wald outlined three elements of screening: the identification of persons likely to be at high risk for a specific disorder so that further testing can be done and preventive actions taken, outreach to populations that have not sought medical attention for the condition, and follow-up and intervention to benefit the screened persons.[4] Several groups have used these principles to develop policies regarding genetic testing in populations.[5] Screening of newborns, which has been carried out in the United States since the early 1960s, serves as a foundation for other types of genetic screening.[6, 7]

NEWBORN SCREENING

Each state (and the District of Columbia) determines its own list of diseases and methods for the screening of newborns. Only phenylketonuria and hypothyroidism are screened for by all these jurisdictions.[7] Table 7.1 lists the disorders that are included in many state programs for newborn screening and gives one an idea of the diversity of techniques employed. The addition of a test or a method to a state's screening program depends on the efforts of advisory boards for newborn screening, political lobbying of legislatures, and the efforts of laboratory personnel for newborn screening. There has often been a lack of research to demonstrate the effectiveness of screening and treatment for a disorder, either before or after the disease is added to the newborn-screening program. The technological spectrum ranges from the original Guthrie bacterial inhibition assay, developed in the late 1950s,[8] to tandem mass spectrometry[9, 10] and DNA analysis.[11–13] With the use of DNA testing of the blood blot obtained from the screening of a newborn, the state of Texas reduced the age at confirmation of the diagnosis of sickle cell disease from four months to two months.[14] Rapid diagnostic confirmation is imperative for the initiation of penicillin prophylaxis to prevent illness and death in patients with sickle cell disease.[15, 16] The cost of this follow-up test is $10 or less for each positive sample from the original screening.[14]

Two-tiered testing is also used for congenital hypothyroidism, since patients with primary hypothyroidism have elevated levels of thyrotropin and low levels

Table 7.1. Disorders Included in Newborn-Screening Programs

Disorder	Screening Method	States Offering Test	Treatment
Phenylketonuria	Guthrie bacterial inhibition assay Fluorescence assay Amino-acid analyzer Tandem mass spectrometry	All	Diet restricting phenylalanine
Congenital hypothyroidism	Measurement of thyroxine and thyrotropin	All	Oral levothyroxine
Hemoglobinopathies	Hemoglobin electrophoresis Isoelectric focusing High-performance liquid chromatography Follow-up DNA analysis	Most	Prophylactic antibiotics Immunization against *Diplococcus pneumoniae* and *Haemophilus influenzae*
Galactosemia	Beutler test Paigen test	Limited no.	Galactose-free diet
Maple syrup urine disease	Guthrie bacterial inhibition assay	Limited no.	Diet restricting intake of branched-chain amino acids
Homocystinuria	Guthrie bacterial inhibition assay	Limited no.	Vitamin B_{12} Diet restricting methionine and supplementing cystine
Biotinidase deficiency	Colorimetric assay	Limited no.	Oral biotin
Congenital adrenal hyperplasia	Radioimmunoassay Enzyme immunoassay	Limited no.	Glucocorticoids Mineralocorticoids Salt
Cystic fibrosis	Immunoreactive trypsinogen assay followed by DNA testing Sweat chloride test	Limited no.	Improved nutrition Management of pulmonary symptoms

of thyroxine.[17, 18] The two-tiered strategy provides better sensitivity and specificity than either test alone. However, the health care professional needs to use clinical judgment in addition to the results of newborn screening. If a patient with a negative newborn-screening test has symptoms of congenital hypothyroidism, clinical acumen should override the test result and specific diagnostic testing should be performed.[17] The results of screening tests are not infallible because of the possibility of biologic, clerical, and laboratory errors.[19–21]

Audiometry is used to screen newborns for hearing defects. The frequency of deafness in childhood is as high as 1 in 500.[22] These programs are based in hospitals and are therefore decentralized.[7] Mutations in the gene for connexin 26 account for 40 percent of all cases of childhood hearing loss, with a carrier rate of 3 percent in the population.[23] A single mutation is responsible for most of these cases in a mixed U.S. population.[23] A different mutation is predominant among Ashkenazi Jews.[24] Two-tiered testing in which audiometry is followed by DNA testing for mutations in the connexin 26 gene may be a useful and cost-effective approach to screening for hearing loss.[25] Early detection provides the possibility of aggressive intervention to improve a child's language skills, provide cochlear implants, or do both.[23]

In 1999, the American Academy of Pediatrics and the Health Resources and Services Administration convened the Newborn Screening Task Force to address the lack of consistency in the disorders included in screening programs and the testing methods used in the various states.[26] The group concluded that there should be a national consensus on the diseases tested for in state programs of newborn screening. The American Academy of Pediatrics, American College of Medical Genetics, Health Resources and Services Administration, Centers for Disease Control and Prevention, March of Dimes, and other groups are working together to create a national agenda for newborn screening.

A disorder that may be included in newborn screening tests is cystic fibrosis. Cystic fibrosis has been included in the newborn-screening program in Colorado since the demonstration that some affected infants had malnutrition as a result of the pancreatic dysfunction.[27] This observation was confirmed by a randomized trial in Wisconsin involving infants with a positive newborn-screening test for cystic fibrosis.[28] In the study, infants with a positive test were randomly assigned to a screened group (in which physicians were informed of the positive screening result) or a control group (in which physicians were informed of the positive screening result when the child was four years of age if cystic fibrosis had not been diagnosed clinically or if the child's parents had not asked about the results of the screening test). In Wisconsin, infants are first tested with the use of an immunoreactive trypsinogen assay;[29] if the result is positive, the test is followed up with a DNA test of the original specimen of dried blood obtained for newborn screening.[30–32] The cost of each follow-up DNA test for infants with positive results on the immunoreactive trypsinogen assay was estimated to be $3 to $5.[31]

A new form of technology, tandem mass spectrometry, detects more than 20 disorders, not all of which can be treated. A justification for introducing tandem mass spectrometry is the identification of newborns with medium-chain

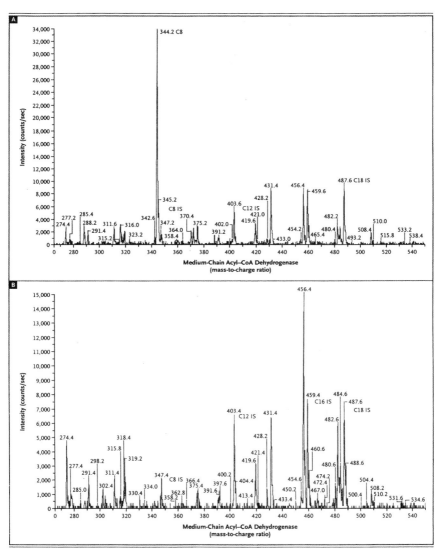

Figure 7.1. Results of Screening by Tandem Mass Spectrometry for Medium-Chain Acyl-Coenzyme A (CoA) Dehydrogenase Deficiency in an Affected Patient (Panel A) and a Control Subject (Panel B). IS denotes internal standard. (Figure provided courtesy of John E. Sherwin, Ph.D.)

acyl-coenzyme A (CoA) dehydrogenase deficiency (Fig. 7.1). Without early detection and intervention, this deficiency leads to episodic hypoglycemia, seizures, coma associated with intercurrent illnesses and fasting, and a risk of death of approximately 20 percent after the first episode in the first and second year of life.[33, 34] Management of medium-chain acyl-CoA dehydrogenase deficiency involves educating families about the dangers of hypoglycemia, which can be triggered by fasting, with resulting fat catabolism, during intercurrent illnesses and by inadequate caloric intake, and of the need for aggressive intervention with intravenous glucose if hypoglycemia does occur. For many of the other disorders detected by tandem mass spectrometry, treatment is not available, but families will potentially be spared "diagnostic odysseys" with a severely ill child.[35] The eventual goal is collaborative research to determine the appropriate treatment after early diagnosis.[7, 36] In addition, this information may be useful for genetic counseling of these families. A cause for concern is that tandem mass spectrometry may detect metabolic variations of unknown clinical significance, creating unwarranted anxiety in parents and health care professionals.

CARRIER SCREENING OF ADULT POPULATIONS FOR SINGLE-GENE DISORDERS

Tay-Sachs Disease

Carrier screening for Tay-Sachs disease has targeted Ashkenazi Jewish populations of childbearing age.[37] In a 30-year period, 51,000 carriers have been identified, resulting in the identification of 1,400 two-carrier couples.[37] Another approach has been taken in Montreal, where high-school students learn about Tay-Sachs disease and thalassemia as part of a biology course. Those of Ashkenazi Jewish descent can request carrier testing for Tay-Sachs disease, and those of Mediterranean ancestry can be tested for thalassemia.[38] When women who have been identified as carriers in high school later consider becoming pregnant, they bring their partners in for testing. Although this program has been very successful in Canada, the culture and the legal environment in the United States, including a standard that does not allow high-school students to consent to medical care and the implications for insurability, may prohibit the adoption of such a model.[39]

Cystic Fibrosis

Northern Europeans have a carrier frequency of cystic fibrosis of 1 in 25 to 1 in 30; the rate is lower in other ethnic and cultural groups.[17] A 1997 National

Institutes of Health Consensus Development Conference[40] recommended that the following populations be screened for mutations associated with cystic fibrosis: the adult family members of patients with cystic fibrosis, the partners of patients with cystic fibrosis, couples planning a pregnancy, and couples seeking prenatal care. Since more than 900 different mutations associated with cystic fibrosis have been reported in the literature,[41] the establishment of screening programs has been difficult. However, the American College of Medical Genetics, the American College of Obstetricians and Gynecologists, and the National Institutes of Health agreed that mutations with a carrier frequency of at least 0.1 percent in the general population should be screened for, resulting in a panel of 25 mutations recommended for carrier testing.[42] These guidelines suggest that carrier testing should be offered to all non-Jewish white persons and Ashkenazi Jews and that other ethnic and cultural groups should be informed of the limitations of the panel to detect carriers in their group (in the case of black persons) or of the low incidence of cystic fibrosis in their group (in the cases of Asian and Native American persons).

Mutations in the gene associated with cystic fibrosis have also been associated with obstructive azoospermia in men[43] and with chronic rhinosinusitis.[44, 45] The guidelines recommend including in the screening panel a test for the R117H mutation, which is associated with congenital bilateral absence of the vas deferens.[42] If the R117H mutation is found, further testing and genetic counseling are recommended.[42]

POPULATION SCREENING FOR
GENETIC SUSCEPTIBILITY TO COMMON DISEASES

Several groups have recently addressed the value of population screening for genetic susceptibility to conditions with onset in adulthood.[46–48] Table 7.2 presents a synthesis of the suggested modifications to the 1968 criteria,[3] based on current principles.

Hereditary hemochromatosis and the thrombophilia that results from carrying a single copy of a factor V Leiden gene are two adult-onset illnesses to which the suggested revised principles for population screening would apply (Table 7.2), and these illnesses also reflect the complex scientific and social issues involved in screening for risk factors for disease. As shown by Wald et al.[49] screening for risk factors for nondiscrete traits that are distributed continuously may not be beneficial even if the factors are associated with a high risk of disease (e.g., high cholesterol levels and heart disease). This is because risk factors are determined by comparing the probability of disease at each

Table 7.2.   Principles of Population Screening as Applied to Genetic Susceptibility to Disease*

Public health assessment

The disease or condition should be an important public health burden to the target population in terms of illness, disability, and death.

The prevalence of the genetic trait in the target population and the burden of disease attributable to it should be known.

The natural history of the condition, from susceptibility to latent disease to overt disease, should be adequately understood.

Evaluation of tests and interventions

Data should be available on the positive and negative predictive values of the test with respect to a disease or condition in the target population.

The safety and effectiveness of the test and accompanying interventions should be established.

Policy development and screening implementation

Consensus regarding the appropriateness of screening and interventions for people with positive and negative test results should be based on scientific evidence.

Screening should be acceptable to the target population.

Facilities should be available for adequate surveillance, prevention, treatment, education, counseling, and social support.

Screening should be a continual process, including pilot programs, evaluation of laboratory quality and health services, evaluation of the effect of screening, and provisions for changes on the basis of new evidence.

The cost effectiveness of screening should be established.

Screening and interventions of screening should be accessible to the target population.

There should be safeguards to ensure that informed consent is obtained and the privacy of those tested is respected, that there is no coercion or manipulation, and that those tested are protected against stigmatization and discrimination.

*Principles are based on Wilson and Jungner,[3] Goel,[46] Khoury et al.,[47] and Burke et al.[48]

end of the distribution of the risk factor (those with the highest level of risk and those with the lowest level of risk). Those with a moderate level of risk are not considered. The likelihood of a disorder, given a positive screening result, is expressed relative to the average risk of the entire population. The goal of screening is to identify individual persons with a high risk in comparison to everyone else.

Hereditary Hemochromatosis

Many consider hereditary hemochromatosis to be the key example of the need for population screening in the genomic era,[50] but gaps in our knowledge preclude the recommendation of population screening for this disorder. This policy issue was discussed by an expert-panel workshop held by the

Centers for Disease Control and Prevention and the National Human Genome Research Institute.[51] The panel concluded that population genetic testing for mutations in HFE, the gene for hereditary hemochromatosis, could not be recommended because of uncertainty about the natural history of the disease, age-related penetrance, optimal care for persons without symptoms who are found to carry mutations, and the psychosocial impact of genetic testing.[52, 53] On the other hand, mutation analysis may be useful in confirming the diagnosis of hereditary hemochromatosis in persons with abnormal indexes of iron metabolism. A meta-analysis of studies[54] showed that homozygosity for the C282Y mutation was associated with the highest risk of hereditary hemochromatosis. The risks associated with other genotypes, including C282Y/H63D and H63DI/H63D, were much lower. A recent large cohort study in the Kaiser Permanente Southern California health care network suggests that the disease penetrance for HFE mutations may be quite low.[55] Only 1 of the 152 subjects who were homozygous for C282Y had symptoms of hereditary hemochromatosis.

Several questions remain regarding the benefits and risks of identifying and treating persons without symptoms who are at high risk for hereditary hemochromatosis (i.e., through population screening). This process should be clearly distinguished from early case finding, which could include testing of iron status, and analysis for mutations in HFE, in persons who present with clinical symptoms consistent with a diagnosis of hereditary hemochromatosis. The natural history of hereditary hemochromatosis—particularly age-related penetrance—remains unknown. Despite the relatively high prevalence of the two most common mutations in the U.S. population,[56] questions persist regarding the nature and prevalence of mutations in specific ethnic and cultural groups, as well as the morbidity[57] and mortality[58] associated with this disease. Therefore, questions remain concerning the persons most likely to benefit from early treatment and thus about the optimal timing of screening and effective intervention, as well as ethical and psychosocial issues[59] (Table 7.2).

Factor V Leiden

Factor V is an important component of the coagulation cascade leading to the conversion of prothrombin into thrombin and the formation of clots.[60] In factor V Leiden, the triplet coding for arginine (CGA) at codon 506 is replaced by CAA, which codes for glutamine (R506Q), resulting in thrombophilia or an increased propensity for clot formation.[61] The prevalence of factor V Leiden varies.[62, 63] Among persons of northern European descent, the prevalence is about 5 percent. The highest prevalence of factor V Leiden is found in Sweden and in some Middle Eastern countries; it is virtually absent in African and Asian populations. Heterozygosity for factor V Leiden results in an increase in the incidence of venous thrombosis by a factor of 4 to 9.[64, 65]

An interaction between factor V Leiden and the use of oral contraceptives was originally found in a case-control study of risk factors for venous thrombosis.[66] Although the use of oral contraceptives alone increases the risk of venous thrombosis by a factor of about 4 and the presence of factor V Leiden alone increases the risk by a factor of about 7, their joint effect was an increase by a factor of more than 30. In spite of the high relative risk, the absolute risk was relatively low (about 28 per 10,000 person-years) among women with factor V Leiden who used oral contraceptives, because the incidence of this complication is relatively low in the population.

The question of whether it is beneficial to screen women for factor V Leiden before prescribing oral contraceptives remains controversial. Venous thrombosis is relatively rare, and the mortality associated with venous thrombosis is low among young women.[67] More than half a million women would need to be screened for factor V Leiden—resulting in tens of thousands of women being denied prescriptions for oral contraceptives—to prevent a single death. In addition to medical and financial considerations, there are issues related to the quality of life, the risk of illness and death from unwanted pregnancy, and concern about possible discrimination by insurance companies. In 2001, the American College of Medical Genetics stated that the opinions and practices regarding testing for factor V Leiden vary considerably, and no consensus has emerged.[68]

For the individual healthy woman contemplating the use of oral contraceptives, the risk-benefit equation does not currently favor screening. For women without symptoms who have family histories of multiple thrombosis, there are no evidence-based guidelines, and decisions will have to be reached individually, without reliance on population-based recommendations.

These examples show why it is essential that data continue to be analyzed to inform decision making for individual persons and populations.

ETHICAL, LEGAL, AND SOCIAL ISSUES

The following are among the ethical, legal, and social issues involved in population-based screening that confront health care providers, policymakers, and consumers.

Testing Children for Adult-Onset Disorders

Two committees of the American Academy of Pediatrics have recently addressed the issue of molecular genetic testing of children and adolescents for adult-onset disease.[69, 70] The Committee on Genetics[69] recommended that persons under 18 years of age be tested only if testing offers immediate medical benefits or if another family member benefits and there is no anticipated harm to the person being tested. The committee regarded genetic counseling before and after testing as an essential part of the process.

The Committee on Bioethics[70] agreed with the Newborn Screening Task Force[27] that the inclusion of tests in the newborn-screening battery should be based on evidence and that there should be informed consent for newborn screening (which is currently not required in the majority of states). The Committee on Bioethics did not support the use of carrier screening in persons under 18 years of age, except in the case of an adolescent who is pregnant or is planning a pregnancy. It recommended against predictive testing for adult-onset disorders in persons under 18 years.

Unanticipated Information

Misattribution of Paternity

The American Society of Human Genetics has recommended that family members not be informed of misattributed paternity unless determination of paternity was the purpose of the test.[71] However, it must be recognized that such a policy may lead to misinformation regarding genetic risk.

Unexpected Associations among Diseases

In the course of screening for one disease, information regarding another disease may be discovered. Although the person may have requested screening for the first disorder, the presence of the second disorder may be unanticipated and may lead to stigmatization and discrimination on the part of insurance companies and employers. Informed consent should include cautions regarding unexpected findings from the testing.

Oversight and Policy Issues

In 1999, the Secretary's Advisory Committee On Genetic Testing was established to advise the Department of Health and Human Services on the medical, scientific, ethical, legal, and social issues raised by the development and use of genetic tests (http://www.od.nih.gov/oba/sacgt.htm).[72] The committee conducted public outreach to identify issues regarding genetic testing. There was an overwhelming concern on the part of the public regarding discrimination in employment and insurance. The advisory committee recommended the support of legislation preventing discrimination on the basis of genetic information and increased oversight of genetic testing. The Food and Drug Administration was charged as the lead agency and was urged to take an innovative approach and consult experts outside the agency. The goal is to generate specific language for the labeling of genetic tests, much as drugs are described in the *Physicians' Desk Reference*.[73] Such labeling would provide persons considering, and health professionals recommending, genetic tests with information about the clinical validity and value of the test—what information the test will provide, what choices will be available to people after they know their test results, and the limits of the test.

In conclusion, although the use of genetic information for population screening has great potential, much careful research must be done to ensure that such screening tests, once introduced, will be beneficial and cost effective.

NOTES

1. Juengst ET. "Prevention" and the goals of genetic medicine. Hum Gene Ther 1995; 6:1595–605.

2. Ransohoff DF, Sandier RS. Screening for colorectal cancer. N Engl JMed 2002; 346:40–4.

3. Wilson JMG, Jungner G. Principles and practice of screening for disease. Public health papers no.34. Geneva: World Health Organization, 1968.

4. Wald NJ. The definition of screening. J Med Screen 2001;8:1.

5. Wilfond BS, Tbornson EJ. Models of public health genetics policy development. In: Khoury MJ, Burke W, Thomson EJ, eds. Genetics and public health in the 21st century: using genetic information to improve health and prevent disease. New York: Oxford University Press, 2000:61–82.

6. Committee for the Study of Inborn Errors of Metabolism. Genetic screening: programs, principles, and research. Washington, D.C.: National Academy of Sciences, 1975.

7. McCabe LL, Therrell BL Jr, McCabe ERB. Newborn screening: rationale for a comprehensive, fully integrated public health system. Mol Genet Metab (in press).

8. Guthrie R, Susi A. A simple phenylalanine method for detecting phenylketonuria in large populations of newborn infants. Pediatrics 1963;32:338–43.

9. Chace DH, Hillman SL, Van Hove JL, Naylor EW. Rapid diagnosis of MCAD deficiency: quantitative analysis of octanoylcar-nitine and other acyl-camitines in newborn blood spots by tandem mass spectrometry. ClinChem 1997;43:2106–13.

10. Andresen BS, Dobrowolski SF, O'Reilly L, et al. Medium-chain acyl-CoA dehydrogenase (MCAD) mutations identified by MS/MS-based prospective screening of newborns differ from those observed in patients with clinical symptoms: identification and characterization of a new, prevalent mutation that results in mild MCAD deficiency. Am J Hum Genet 2001;68:1408–18.

11. McCabe ERB, Huang S-Z, Seltzer WK, Law ML. DNA microextraction from dried blood spots on filter paper blotters: potential applications to newborn screening. Hum Genet 1987;75:213–6.

12. Links DC, Minter M, Tarver DA, Vanderford M, Hejtmancik JF, McCabe ERB. Molecular genetic diagnosis of sickle cell disease using dried blood specimens on blotters used for newborn screening. Hum Genet 1989;81:363–6.

13. Descartes M, Huang Y, Zhang Y-H, et al. Genotypic confirmation from the original dried blood specimens in a neonatal hemoglobinopathy screening program. Pediatr Res 1992;31:217–21.

14. Zhang Y-H, McCabe LL, Wilborn M, Therrell EL Jr, McCabe ERB. Application of molecular genetics in public health: improved follow-up in a neonatal hemoglobinopathy screening program. Biochem Med Metab Biol 1994;52:27–35.

15. Gaston MH, Verter JI, Woods G, et al. Prophylaxis with oral penicillin in children with sickle cell anemia: a randomized trial. N Engl J Med 1986; 314:1593–9.

16. Consensus Development Panel. Newborn screening for sickle cell disease and other hemoglobinopathies. NIH consensus statement Vol. 6. No. 9. Bethesda, Md.: NIH Office of Medical Applications of Research, 1987:1–22.

17. American Academy of Pediatrics Committee on Genetics. Newborn screening fact sheets. Pediatrics 1989;83:449–64.

18. Burrow GN, Dussault JH, eds. Neonatal thyroid screening. New York: Raven Press, 1980:155.

19. McCabe ERB, McCabe L, Mosher GA, Allen RJ, Berman IL. Newborn screening for phenylketonuria: predictive validity as a function of age. Pediatrics 1983; 72:390–8.

20. Holtzman C, Slazyk WE, Cordero JF, Hannon WH. Descriptive epidemiology of missed cases of phenylketonuria and congenital hypothyroidism. Pediatrics 1986;78:553–8.

21. Dequeker E, Cassiman J-J. Quality evaluation of data interpretation and reporting. Am J Hum Genet 2001;69:Suppl:438. abstract.

22. Mehl AL, Thomson V. The Colorado Newborn Hearing Screening Project, 1992–1999: on the threshold of effective population-based universal newborn hearing screening. Pediatrics 2002;109:134. abstract.

23. Cohn ES, Kelley PM. Clinical phenotype and mutations in connexin 26 (DFNB1/GJB2), the most common cause of childhood hearing loss. Am J Med Genet 1999;89:130–6.

24. Morrell RJ, Kim HL, Hood LJ, et al. Mutations in the connexin 26 gene (GJB2) among Ashkenazi Jews with nonsyndromic recessive deafness. N Engl J Med 1998;339: 1500–5.

25. McCabe ERB, McCabe LL. State-of-the-art for DNA technology in newborn screening. Acta Paediatr Suppl 1999;88:58–60.

26. Newborn Screening Task Force. Serving the family from birth to the medical home: newborn screening: a blueprint for the future—a call for a national agenda on state newborn screening programs. Pediatrics 2000;106:389–422.

27. Reardon MC, Hammond KB, Accurso FJ, et al. Nutritional deficits exist before 2 months of age in some infants with cystic fibrosis identified by screening test. J Pediatr 1984;105:271–4.

28. Farrell PM, Kosorok MR, Rock MJ, et al. Early diagnosis of cystic fibrosis through neonatal screening prevents severe malnutrition and improves long-term growth. Pediatrics 2001;107:1–13.

29. Hassemer DJ, Laessig RH, Hoffman GL, Farrell PM. Laboratory quality control issues related to screening newborns for cystic fibrosis using immunoreactive trypsin. Pediatr Pulmonol Suppl 1991;7:76–83.

30. Seltzer WK, Accurso F, Fall MZ, et al. Screening for cystic fibrosis: feasibility of molecular genetic analysis of dried blood specimens. Biochem Med Metab Biol 1991;46:105–9.

31. Gregg RG, Wilfond BS, Farrell PM, Laxova A, Hassemer D, Mischler EH. Application of DNA analysis in a population-screening program for neonatal diagnosis of cystic fibrosis (CF): comparison of screening protocols. Am J Hum Genet 1993;52:616–26.

32. Kant JA, Mifflin TE, McGlennen R, Rice E, Naylor E, Cooper DL. Molecular diagnosis of cystic fibrosis. Clin Lab Med 1995;15: 877–98.

33. Roe CR, Ding J. Mitochondrial fatty acid oxidation disorders. In: Scriver CR, Beaudet AL, Sly WS, Valle D, eds. The metabolic & molecular bases of inherited disease. 8th ed. Vol. 2. New York: McGraw-Hill, 2001:2297–326.

34. Matsubara Y, Narisawa K, Tada K, et al. Prevalence of K329E mutation in medium-chain acyl-CoA dehydrogenase gene determined from Guthrie cards. Lancet 1991;338:552–3.

35. Wilcken B, Travert G. Neonatal screening for cystic fibrosis: present and future. Acta Paediatr Suppl 1999;88:33–5.

36. Naylor EW, Chace DH. Automated tandem mass spectrometry for mass newborn screening for disorders in fatty acid, organic acid, and amino acid metabolism. J Child Neurol 1999;14:Suppl 1:S4–S8.

37. Kaback MM. Population-based genetic screening for reproductive counseling: the Tay-Sachs disease model. EurJ Pediatr 2000;159:Suppl 3:S192–S195.

38. Mitchell JJ, Capua A, Clow C, Scriver CR. Twenty-year outcome analysis of genetic screening programs for Tay-Sachs and thalassemia disease carriers in high schools. Am J Hum Genet 1996;59:793–8.

39. McCabe L. Efficacy of a targeted genetic screening program for adolescents. Am J Hum Genet 1996:59:762–3.

40. Genetic testing for cystic fibrosis: National Institutes of Health Consensus Development Conference statement on genetic testing for cystic fibrosis. Arch Inter Med 1999;159:1529–39.

41. Grody WW, Desnick RJ. Cystic fibrosis population carrier screening: here at last—are we ready? Genet Med 2001;3:87–90.

42. Grody WW, Cutting GR, Klinger KW, Chards CS, Watson MS, Desnick RJ. Laboratory standards and guidelines for population-based cystic fibrosis carrier screening. Genet Med 2001;3:149–54.

43. Mak V, Zielenski J, Tsui L-C, et al. Proportion of cystic fibrosis gene mutations not detected by routine testing in men with obstructive azoospermia. JAMA 1999;281: 217–24.

44. Raman V, Clary R, Siegrist KL, Zehnbauer B, Chatila TA. Increased prevalence of mutations in the cystic fibrosis transmembrane conductance regulator in children with chronic rhinosinusitis. Pediatrics 2002;109:136–7. abstract.

45. Wang XJ, Moylan B, Leopold DA, et al. Mutation in the gene responsible for cystic fibrosis and predisposition to chronic rhinosinusitis in the general population. JAMA 2000;284:1814–9.

46. Goel V. Appraising organized screening programmes for testing for genetic susceptibility to cancer. BMJ 2001;322:1174–8.

47. Khoury MJ, Burke W, Thomson EJ. Genetics and public health: a framework for the integration of human genetics into public health practices. In: Khoury MJ, Burke W, Thomson EJ, eds. Genetics and public health in the 21st century: using genetic information to improve health and prevent disease. New York: Oxford University Press, 2000:3–24.

48. Burke W, Coughlin SS, Lee NC, Weed DL, Khoury MJ. Application of population screening principles to genetic screening for adult-onset conditions. Genet Test 2001;5:201–11.

49. Wald NJ, Hackshaw AK, Frost CD. When can a risk factor be used as a worthwhile screening test? BMJ 1999;319:1562–5.

50. Collins FS. Keynote speech at the Second National Conference on Genetics and Public Health, December 1999. Atlanta: Office of Genetics & Disease Prevention, 2000. (Accessed December 6, 2002, at http://www.cdc.gov/genomics/info/conference/intro. htm.)

51. Cogswell ME, Burke W, McDonnell SM, Franks AL. Screening for hemochromatosis: a public health perspective. Am J Prev Med 1999;16:134–40.

52. Burke W, Thomson E, Khoury MJ, et al. Hereditary hemochromatosis: gene discovery and its implications for population-based screening. JAMA 1998;280:172–8.

53. EASL International Consensus Conference on Hemochromatosis. III. Jury document. J Hepatol 2000;33:96–504.

54. Burke W, Imperatore G, McDonnell SM, Baron RC, Khoury MJ. Contribution of different HFE genotypes to iron overload disease: a pooled analysis. Genet Med 2000;2:271–7.

55. Beuder E, Felitti VJ, Koziol JA, Ho NJ, Gelbart T. Penetrance of 845GA (C282Y) HFE hereditary haemochromatosis mutation in the USA. Lancet 2002;359:211–8.

56. Steinberg KK, Cogswell ME, Chang JC, et al. Prevalence of C282Y and H63D mutations in the hemochromatosis (HFE) gene in the United States. JAMA 2001;285:2216–22.

57. Brown AS, Gwinn M, Cogswell ME, Khoury MJ. Hemochromatosis-associated morbidity in the United States: an analysis of the National Hospital Discharge Survey, 1979–1997. Genet Med 2001;3:109–11.

58. Yang Q, McDonnell SM, Khoury MJ, Cono J, Parrish RG. Hemochromatosis-associated mortality in the United States from 1979 to 1992: an analysis of Multiple-Cause Mortality Data. Ann Intern Med 1998;129:946–53.

59. Imperatore G, Valdez R, Burke W. Case study: hereditary hemochromatosis. In: Khoury ML, Little J, Burke W, eds. Human genome epidemiology: scientific foundation for using genetic information to improve health and prevent disease. New York: Oxford University Press (in press).

60. Greenberg DL, Davie EW. Introduction to hemostasis and the vitamin K-dependent coagulation factors. In: Scriver CR, Beaudet AL, Sly WS, Valle D, eds. The metabolic & molecular bases of inherited disease. 8th ed. Vol. 3. New York: McGraw-Hill, 2001:4293–326.

61. Esmon CT. Anticoagulation protein C/thrombomodulin pathway. In: Scriver CR, Beaudet AL, Sly WS, Valle D, eds. The metabolic & molecular bases of inherited disease. 8th ed. Vol. 3. New York: McGraw-Hill, 2001:4327–43.

62. Rees DC, Cox M, Clegg JB. World distribution of factor V Leiden. Lancet 1995;346:1133–4.

63. Ridker PM, Miletich JP, Hennekens CH, Buring JE. Ethnic distribution of factor V Leiden in 4047 men and women: implications for venous thromboembolism screening. JAMA 1997;277:1305–7.

64. Rosendaal FR, Koster T, Vandenbroucke JP, Reitsma PH. High risk of thrombosis in patients homozygous for factor V Leiden (activated protein C resistance). Blood 1995;15:1504–8.

65. Emmerich J, Rosendaal FR, Catraneo M, et al. Combined effect of factor V Leiden and prothrombin 20210A on the risk of venous thromboembolism—pooled analysis of 8 case-controlled studies including 2310 cases and 3204 controls. Thromb Haemost 2001;86:809–16. [Erratum, Thromb Haemost 2001;86:1598.]

66. Vandenbroucke JP, Koster T, Briet E, Reitsma PH, Bertina RM, Rosendaal FR. Increased risk of venous thrombosis in oral-contraceptive users who are carriers of factor V Leiden mutation. Lancet 1994;344:1453–7.

67. Vandenbroucke JP, van der Meer FJM, Helmerhorst FM, Rosendaal FR. Factor V Leiden: should we screen oral contraceptive users and pregnant women? BMJ 1996;313:1127–30.

68. Grody WW, Griffin JH, Taylor AK, Korf BR, Heit JA. American College of Medical Genetics consensus statement on factor V Leiden mutation testing. Genet Med 2001;3:139–48.

69. Committee on Genetics. Molecular genetic testing in pediatric practice: a subject review. Pediatrics 2000;106:1494–7.

70. Nelson RM, Botkjin JR, Kodish ED, et al. Ethical issues with genetic testing in pediatrics. Pediatrics 2001;107:1451–5.

71. The American Society of Human Genetics. Statement on informed consent for genetic research. Am J Hum Genet 1996;59:471–4.

72. McCabe ERB. Clinical genetics: compassion, access, science, and advocacy. Genet Med 2001;3:426–9.

73. Physicians' desk reference. 56th ed. Montvale, N.J.: Medical Economics, 2002.

8

Navigating Race in the Market for Human Gametes

Hawley Fogg-Davis

Since the first successful birth resulting from in vitro fertilization in 1978, ethicists have debated a wide spectrum of moral questions raised by IVF, including concerns about economic exploitation, profiteering, health effects on women's bodies, interference with traditional family norms, and children's welfare.[1] Yet these discussions rarely, if ever, address the racially selective use of reproductive technologies. Legal scholar Dorothy Roberts has documented a racial disparity in access to and use of reproductive technologies, pointing out that even though black women experience infertility at higher rates than white women, white women are twice as likely as black women to use reproductive technologies.[2] But no one has yet explored the production and reproduction of racial meanings *within* this newfangled market.

How do descriptive and prescriptive notions of race affect the economic behavior of those who possess the financial means, time, and cultural capital to pursue assisted reproduction? Conversely, how do the racial choices of gamete consumers shape contemporary notions of race?[3] Are whites, who comprise the overwhelming majority of gamete consumers, morally justified in choosing the gametes of a white donor?[4] Is same-race preference among black or other nonwhite gamete shoppers morally different from same-race preference among whites? Do cross-racial choices, such as a white couple's request for an Asian American egg donor, amount to benign or invidious racial discrimination? In sum, what role, if any, should race play in the selection and purchase of human reproductive tissue?

Race-based gamete selection raises two major, linked ethical issues. One is the harm that racial stereotyping causes to individuals, and the second is the public awareness that racial stereotyping is an accepted feature of this largely unregulated market.[5] Choosing a donor according to racial classification is based on racial stereotypes of what that donor is like, and of what a child produced using that person's gametes will be like, as well as the gamete consumer's own racial self-concept and racial aspirations. Race-gamete selection is tied to race-based desires in family formation. The dangerous subtext, or subliminal message, conveyed by race-based gamete choice is that a child created using the gametes donated by a racially designated person ought to adopt a race-specific cultural disposition; and develop his or her self-concept within those parameters. The net result is the constriction of individual freedom in forging one's identity.

Negative social repercussions also flow from this process of racial sorting. Naomi Zack argues that the white American family has historically been and continues to be "a publicly sanctioned private institution for breeding white people."[6] Race-specific gamete shopping underscores and extends Zack's point.

Assisted reproduction, as the name suggests, brings reproductive decision-making into public view. Racial choices made in this arena publicly reinforce and make explicit the routine use of racial discrimination in the choice of a partner for procreative sexual intercourse. It is not so much that the former is morally worse than the latter. Both operate on the level of racial stereotype, prejudging and weeding out certain individuals based at least partly on their ascribed race.

The unique problem of racial choice in the gamete market lies in *how* interpersonal racial choices are expressed. Noncoital reproduction requires people to articulate a race-based reproductive choice that usually remains unspoken in coital reproduction. The price tag attached to these racial reproductive choices enhances the publicity of the stereotyping.

Explicit racial selectivity in the gamete market has the potential to uncover submerged racial biases that permeate the U.S. social terrain. But if we unearth these racial desires only to ignore them, thereby affirming them by default, then we end up sanctioning stereotypes of race-based familial structure. The fact that racially coded donor profiles exist and can be viewed by the public makes this practice part of our public consciousness. Hence, race-based donor choices are inextricably tied to public notions of the normative role that race ought to play in family formation.

My argument against this mode of racial stereotyping is not based in color blindness or a call for abolishing racial categories. Race can and should be a source of self-identification, and to some extent group identification, but it should never be overwhelming or fixed. What is needed, instead, is a way for individuals to mediate or navigate over the course of their lives between the racial categories ascribed to them and their own racial self-identification.

I call this theoretical concept "racial navigation." Racial navigation recognizes the practical need for individuals living in a race-conscious society to acknowledge the social and political weight of racial categories, while urging individuals to resist passively absorbing these expectations into their self-concepts. My objective is to maximize human freedom under the existential pressure of racial categories. While racial navigation begins at the personal level, I intend for it to guide interpersonal conduct in the market for human gametes and beyond.[7] Before delving into the theoretical underpinnings of racial navigation, and demonstrating how it might mitigate the perpetuation of racial stereotypes in the gamete market, I want first to give a brief overview of how race is marketed in the business of paid gamete donation.

THE RACIAL MARKETING OF HUMAN GAMETES

The U.S. fertility treatment business is a booming, multibillion-dollar industry. With infertility rates on the rise,[8] the number of clinics offering IVF has risen sharply since the mid-1980s to approximately 330 nationwide.[9] Largely unregulated, these clinics compete fiercely with each other for a market of approximately 2.1 million infertile married couples.[10] Only a small percentage of these couples are likely to pursue IVF, donor insemination, or other assisted reproductive technologies.[11] And those who do pursue assisted reproduction have to be wealthy enough to afford fertility services such as IVF, which costs an average of $7,800 per cycle.[12] In discussing her finding that white women are twice as likely as black women to use reproductive technologies,[13] Roberts suggests that the disparity may "stem from a complex interplay of financial barriers, cultural preferences, and more deliberate professional manipulation" (p. 253).

Roberts argues that most blacks de-emphasize the role of genetics in both familial and community membership, as well as in the process of personal identity. Whereas many whites have historically gone to extraordinary and absurd lengths to guard against the "pollution" of a white "blood-line" by either avoiding interracial sexual relationships or evading the consequences of such relationships, Roberts notes a general attitude of acceptance of *mélange* within black families and extended kin networks:

> The notion of racial purity is foreign to Black folks. Our communities, neighborhoods, and families are a rich mixture of languages, accents, and traditions, as well as features, colors, and textures. Black life has a personal and cultural hybrid character. There is often a mélange of physical features—skin and eye color, hair texture, sizes, and shapes—within a single family. We are used to "throwbacks"—a pale, blond child born into a dark-skinned family, who

inherited stray genes from a distant white ancestor. . . . We cannot expect our
children to look just like us (p. 263).

Even though prejudice against dark skin color and "African" physical features has existed within black communities since slavery, Roberts maintains,
"sharing genetic traits seems less critical to Black identity than to white identity" (p. 263).

It is important to distinguish between *genetic traits* and *genetic ties*. Genetic traits refer, in the case of physical race, to the *physical expression* of
genes inherited from biological parents. This genetic inheritance includes recessive genes—genes that are not physically expressed—and this (among
other factors[14]) makes it possible for children to have physical characteristics
that differ from those of their biological parents. A genetic tie, on the other
hand, refers to the simple fact of sharing genes with another person, a biological relative. Although having a genetic tie to someone often means sharing genetic traits, the two are conceptually distinct. You and I may share the
genetic trait of big ears and have no genetic tie. Likewise, genetically tethered
sisters may share very few genetic traits. Roberts is right to point out that
black Americans tend to de-emphasize genetic traits when it comes to determining who is black. Acceptance of *mélange* among African Americans is a
practical response to the prior white American existential claim that any person with a black family member (a black genetic tie, not a black genetic trait)
cannot be a white person.[15]

Genetic ties are another matter. While it is true that blacks have been more
likely than whites to develop kin networks among nongenetically related individuals, it is not clear that contemporary black Americans de-emphasize the
value of genetic ties to the extent Roberts implies.[16] Given the syncretic and
hybrid nature of black cultural practices, which Roberts concedes, particularly the deceptively simple aspiration to be an American,[17] it seems unlikely
that we can exempt blacks entirely from the widespread, culturally based desire to have one's "blood-line" perpetuated *vis à vis* genetically related children.

Racial disparity in the fertility services market more likely stems from economic disparities between blacks and whites, professional manipulation, and
historically based fears of technological intervention with reproduction.[18]
Whatever the reasons for racial disparity among fertility consumers, the fact
remains that most of them are white, middle and upper class, married couples.
The expressed and anticipated demands of these individuals shape the contours of today's gamete market.

Racial category is the primary criterion used by those interested in buying
human eggs and sperm for the purposes of donor insemination and IVF. Race

is also often prominent in private advertisements soliciting egg and sperm donors; it is the first category on the donor lists of most fertility clinics, many of which are publicly accessible via the Internet.[19] Donor lists include various details about the people who have contracted with the clinic to sell their gametes. Objective facts such as blood type, height, weight, and eye color are listed alongside more subjective "facts" such as hair texture, ethnic origin, skin tone, and tanning ability. Self-reported skills, accomplishments, and boasts—like years of education and athletic and musical talent—find a place on the screen or page next to the donor's favorite color, foods, and hobbies. At the California Cryobank in Beverly Hills, a gamete shopper can even judge donors' responses to the quintessential question, "Where do you see yourself in five years?" Donor profiles increasingly bulge with information that ranges from vital health information to genetically irrelevant details such as a donor's self-reported life goals. And the "more is better" trend in donor profile information is likely to continue, in step with the elusive quest for a comprehensive picture of the genetic material with which the consumers of this human tissue will attempt to create a baby.

The California Cryobank offers an online donor catalog where the serious and curious alike can pore over an extensive grid of one hundred and seventy-two sperm donors.[20] Of these donors, 146 are listed as Caucasian, fifteen as Other, nine as Black/African American, and sixteen as Asian. In addition to "racial group," each donor's profile contains an abbreviated statement of "ethnic origin." Most of the Caucasian donors list multiple ethnic origins; many claim three and four different ethnicities. For example, donor 993 describes himself as Caucasian and of Irish, Russian, English, and French descent. All but one of the Asian donors describe themselves as monoethnic—as, for example, Korean or Filipino. Almost every African American donor describes himself as ethnically African American. Two black donors describe themselves as Nigerian, and one self-identifies as African American and Ethiopian. The "racial group" of Other contains an eclectic array of ethnic "mixes"—German and Chinese, Pacific Islander, and African American—as well as the singular "ethnic origins" of Mexican and East Indian. Donors are further subdivided by their "skin tones," which range from fair to medium to olive to dark.

The above donor catalog highlights the idiosyncratic and *ad hoc* use of racial classification in U.S. society. From the very moment we try to place individuals into racial boxes we discover that the center cannot hold and things very rapidly fall apart. Why do members of the Caucasian group typically have multiple ethnic origins while members of the African American/Black racial group almost universally have an ethnic origin equivalent to their racial group? How does one really distinguish between an olive complexioned Caucasian man and an "Other" who describes himself as Italian and African American with fair skin? Should the staff of California Cryobank decide what racial box to check, or should each individual donor have the freedom to describe his racial identity using language that transcends the sperm bank's racial boxes? And how are racial descriptions of sperm donors related to the consumer's goal of creating a baby?

RACE-BASED SOCIAL ONTOLOGY

Answering this last question requires inquiry into the meaning of race in our current social ontology. Charles Mills defines social ontology as "the basic struts and girders of social reality," "analogous to the way 'metaphysics' *simpliciter* refers to the deep structure of reality as a whole."[21] This deep structure is not, in Mills's view, metaphysical. Instead, racial categories are devised, maintained, and revised through political decisions. Mills defines racial constructivism in the following way: "The intersubjectivist agreement in moral and scientific constructivism is a hypothetical agreement of all under *epistemically* idealized conditions. Racial constructivism, by contrast, involves actual agreement of some under conditions where the constraints are not epistemic (getting at the truth) but *political* (establishing and maintaining privilege)" (p. 48). In this sense, race is not metaphysical, but a *"contingently* deep reality that structures our particular social universe, having a social objectivity and causal significance that arise out of *our* particular history" (p. 48).

Such inquiry is not limited to individual acts of racism. Instead, Mills and others point to a more insidious kind of racial hierarchy that has been built over a series of political decisions. Structural racism refers to official and unofficial social policies that invidiously affect the lives of nonwhites but cannot be traced to the actions of specific individuals.[22] I do not share Derrick Bell's pessimism that structural racism is permanent.[23] Indeed, the contours of U.S. structural racism have changed over time from chattel slavery to *de jure* segregation to our present circumstance, and this last stop will not be our final destination. As Michael Banton reminds us, the meaning of race has shifted significantly over the last three centuries throughout the globe, and will continue to change in the future.[24]

But change is slow. Racial classification continues to be a source of social hierarchy, a mark of civic standing, cultural development, beauty, intelligence, and subordination. All of us engage this drama. The weight of structural racism on individual lives is felt in the memoir of Toi Derricotte, a light skinned black woman who is often perceived as white. Derricotte explores her own racism against darker skinned blacks, and her action and inaction in the face of racist comments from white neighbors, colleagues, and cab driv- ers and others who believe her to be one of "them."[25] Proof that white skin continues to expand one's social and economic opportunities is brought into sharp focus by Derricotte's experience of shopping for a house in a wealthy and predominately white suburb of New York City. "I had decided not to take my husband with me to the real estate offices because when I had, since he is recognizably black, we had been shown houses in entirely different neigh- borhoods, mostly all-black. . . . At night, under cover of darkness, I would take him back to circle the houses that I had seen and I would describe the in- sides" (p. 13).

As a form of structural racism, housing discrimination has been resistant to antidiscrimination policies and law, which the Supreme Court has interpreted to require evidence of discriminatory intent or purpose. Ultimately, housing discrimination supports a social value that many Americans subscribe to, but rarely express out loud: the right to live in a race-specific neighborhood, a preference that is often translated into the economic right to maintain one's property value. Where one lives greatly affects one's social status, as does the racial composition of one's family portrait; the two are connected, as Derri- cotte's decision not to introduce her husband to the real estate agent illustrates. Though racial discrimination in housing and gamete markets are different in many respects, both imply a greater degree of intimacy than pub- lic accommodations such as hotels, theatres, and restaurants. There is a strong presumption in favor of individual autonomy when it comes to decisions that affect who one must interact with on the home front.

John Robertson argues for an expansive notion of what he terms "procre- ative liberty" in the market for reproductive technologies. On Robertson's view, "individuals should be free to use these techniques or not as they choose, without government restriction, unless strong justification for limit- ing them can be established" (p. 4). Such justification is "seldom present," and is limited to preventing women from using their reproductive capacity for nonreproductive ends such as producing fetal tissue for research and trans- plant.[26] He discusses the ethical conflicts arising from "quality control measures" that use technologies to screen out and select for genetic charac- teristics, but he avoids the subject of race completely. Ironically, the follow- ing statement could support my concern about racial stereotyping if Robert- son considered race-based procreative choices a kind of "quality control": "Quality control measures may in practice not be optional for many women, and may place unrealistic expectations on children who are born after prena- tal screening" (p. 11).

I agree with Robertson that government should not restrict racial discrimi- nation in the gamete market, but disagree with his decision to shield these choices from moral investigation. Concern for individual freedom should motivate inquiry into the ways that a race-driven market in human reproductive tissue is likely to constrain personal identity expression. Robertson fails to acknowledge this possibility because he restricts his notion of individual liberty to the freedom to make procreative choices in a free economic market. As with many libertarian arguments, the status quo becomes ground zero, and little or no attempt is made to dig below its surface.[27]

Ground zero consists of a socially diffuse system of racial classification that threatens to trap individuals in racial stereotypes. Like gamete shoppers who create the demand for racially labeled donors, the suburban real estate agent automatically associated Derricotte's white phenotype with a certain set of cultural practices and behaviors that were then aligned with racially coded neighborhoods. Robin Kelley and others have criticized this tendency to treat "culture" and "behavior" as synonyms.[28] Culture describes a set of available practices and artifacts that have evolved over time and will continue to change. Individuals respond to cultural menus in different ways, and they should be encouraged in this personal expression.[29] Individual cultural choices and personal behavior among residents of predominately black urban neighborhoods, for example, are not monolithic. Kelley reminds us that "By conceiving black urban culture in the singular, interpreters unwittingly reduce their subjects to cardboard typologies who fit neatly into their own definition of the 'underclass' and render invisible a wide array of complex cultural forms."[30]

NAVIGATING BETWEEN RACIAL IMPOSITION AND RACIAL SELF-IDENTIFICATION

It is this varied response to the imposition of racial classification that is missing from the racial menus of fertility clinics trading in gametes. If consumer demand is rooted solely in a visual, third person conception of race, then the California Cryobank and other fertility clinics will tailor their business strategies to satisfy that racial demand. But racial identification also involves a cognitive dimension, as Robert Gooding-Williams points out.[31] This cognitive dimension creates psychological space for the ongoing process of racial navigation. Racial navigation describes the activity of fending off simplistic and rigid notions of racial identity both in one's self-understanding and in the perception of others. It is a normative theoretical tool available to all living in a system of racial classification. Racial navigation recognizes the practical and strategic need to make sense of oneself within a social ontology of racial categories, to see oneself through the eyes of others in order to challenge that imposition and create new racial meanings for oneself.[32] There is no endpoint for racial navigation. The goal is to create and sustain a fluid self-concept that recognizes the existential weight of racial categories but does not accept them as adequate descriptions of human beings.

Again, racial navigation is made possible by the cognitive dimension of racial meaning. A person may look white, but know herself to be black based on the social convention that anyone with one black ancestor is classified as black. The seeming paradox of being a "white black woman," as law professor Judy Scales-Trent describes herself, is made possible by this rule of hypodescent.[33] Gooding-Williams illuminates this notion by distinguishing between *being black* and *being a black person*.[34] Being black is a third person identification that entails being classified by others as visually black or cognitively black—that is, black according to the social convention of hypodescent. Being a black person, on the other hand, is a first person identity that refers to a person's decision to navigate between these two levels of personal and social meaning, to "make choices, to formulate plans, to express concerns, etc. in light of one's identification of oneself as black" (p. 23). When a black person passes for white she understands herself to be engaged in an act of (willful or unintentional) deception, and everyone knows that such deception is possible.[35]

So being seen or thought of as black, according to the social rule of hypodescent, is "a necessary but not sufficient condition of being a black person" (p. 58). To become a black person one must actively incorporate the fact that one has been designated as black into one's self-concept. A person who is visibly and cognitively designated as white, but who decides to affect certain stereotypical black cultural practices, thus thinking of himself as a black person, can never become a black person since he has failed the first criterion of being socially "seen" as black. The derogatory term "whigger," popular in the mid-1990s, conveys the artificiality, even offensiveness, of the white suburban youth who listens to rap music incessantly and mimics black urban slang.[36] "Whiggers" do not challenge racial stereotypes because their participation in black cultural forms is typically fleeting and no one believes them to be "really" black. Individuals who do satisfy the first criterion of being black can either absorb the racial expectations of others (stereotypes) or challenge the flatness of racial imposition by personalizing their black identities. As Gooding-Williams notes, those identified by others as being black can and often do express their black personhood in an infinite number of ways.

BARRIERS TO RACIAL NAVIGATION IN THE GAMETE MARKET

It is not impossible for racial navigation to begin after racially coded gametes have been bought. But the unexamined use of race to choose gamete donors makes it less likely that people will question third person racial meanings in their self-concepts, family interactions, and social behavior. And even if racial navigation is jump-started after the point of purchase, there is still the lingering damage of the initial racial restriction imposed on market actors, as well as the race-based expectations for children born using donated gametes. Racially organized gamete markets will have profound negative personal and social consequences even if the participants start to navigate racial meanings after they exit the market.

First person views of race fall out of the gamete market altogether. Sperm and egg donors are classified according to a third person view of race for the specific purpose of satisfying race-specific consumer demand. A fertility clinic's business success depends on its staff seeing a prospective donor's racial classification through the eyes of actual and potential customers. At the California Cryobank a client can pay an additional fee for the services of a matching counselor who tries to make an even more precise "match" between the genetic traits desired by the consumer and the genetic traits of particular donors. The PBS Frontline documentary, "Making Babies," showed a matching counselor scrutinizing a sperm donor's photograph while describing his physical features to a client over the telephone.

When counselors examine a donor's profile they find a third person account of that donor's racial identification. They must then shift perspective and attempt to see the donor's racial identification and more pointedly his racial traits through the eyes of the shopper with his or her racial desires. The result is a kind of third person racial identification "once removed." During the economic transaction, the donor's first person sense of racial self recedes further and further into the background, as the idea of race becomes packaged as a genetic commodity that can be detached from particular persons for the purpose of economic trade. In turning race into a genetic commodity, these market forces obliterate, or at least seriously erode, the donor's first person sense of racial self. The absence of first person expression of racial identity is critical because first person views mitigate the binding effects of racial classification, and serve as a break on our will to confine people to racial boxes.

One might argue that first person accounts of racial identity are present in the gamete market in the form of the copious "personal" information contained in donor profiles. The demand for extensive donor profiles might signal that consumers want to know what *sort* of Korean-American a sperm donor is, that they are interested in looking beyond the first column of donors' ascribed race. This is a step toward racial navigation, but the additional information in these donor files is too general to establish the robust sense of first person, raced identity that I have in mind. Donors respond to questions like whether they are athletic, what their career goals entail, their favorite color, and such. They are not asked to give an account of how they have responded to the imposition of race. I am not suggesting that this question should be added to donor questionnaires. The enormity of such an existential question, and the expectation that one's answer will change over time, makes it impossible to answer in a paragraph on a form. I am suggesting, however, that gamete shoppers consider the motivation behind their race-based choices for sperm and ova in the first place.

MOTIVATIONS AND CONSEQUENCES

Whites overwhelmingly demand "white" gametes, as evidenced by the fact that 85 percent of the donors "hired" by California Cryobank are identified as "Caucasian." When people buy gametes according to the racial classification of the donor, they are saying that race is heritable and relevant to their vision of family structure. They are saying that a person's gametes are transmitters of racial meaning that can and should be selectively transmitted to "their" child, through the use of reproductive technologies.[37] The idea that race is a bundle of heritable characteristics such that "all members of these races share certain traits and tendencies with each other that they do not share with members of any other race" is what K. Anthony Appiah calls *racialism*.[38] The racial classification of gametes in the fertility market can be described as *genetic* racialism—the false belief that human genetic tissue transmits specific racial traits and tendencies to future human beings.

So far I haven't said much about the difference between racialism and racism. I've discussed the infusion of our social ontology with a powerful system of racial classification, and I've ventured into the controversial territory of structural racism, suggesting that these two concepts frame the individual actions of gamete shoppers. But I need to address the question of individual racial motivation more directly. Are the consumer demands of gamete shoppers racist? Racism connotes intent to harm someone or some group of people based solely or primarily on racial classification. Do white gamete shoppers *intend* to inflict racially invidious harm when they satisfy their desire for a white family through the purchase of a white donor's gametes? Are nonwhite gamete shoppers racist when they choose gametes from donors who share their own racial classification? What of the person who picks a donor of a racial category different from his own?

Appiah draws a distinction between racialism and racism. For him, racialism is false but not necessarily racist. Racialism is "a cognitive rather than a moral problem" (p. 13). In order for racialism to be racist the racial classifier must attach some moral significance to racial demarcation. But Appiah's position errs in two respects. First, the distinction between racialism and racism is not very helpful in excavating the morality of racial choice in the gamete market because there is no reliable way of gauging whether or not a racial choice has become a "moral problem." The distinction relies on a bright line between motivation and consequence that does not make much of a moral difference. Racialism, no matter how well intentioned, is always poised to cause moral difficulty in the form of racial stereotyping. This constant threat is precisely why racial navigation calls for constant vigilance.

The second problem concerns Appiah's claim that racialism is false, and that racial categories should be abolished because of the restrictions they exert on personal lives: "It is not that there is *one* way that blacks should behave, but that there are proper black modes of behavior. These notions provide loose norms or models, which play a role in shaping the life plans of those who make these collective identities central to their individual identities: of the identities of those who fly under these banners."[39] Though Appiah recognizes and appreciates the recuperative value that racial self-identification can have in the form of, for instance, black power counter-narratives, he urges us to move beyond racialism as a long-term goal (p. 614). Based on this reasoning, I suspect Appiah would advocate the removal of racial classification from gamete donor lists altogether.

I agree with Appiah that racialism places restrictions on the expression and life plans of individuals. Indeed, my notion of racial navigation is consistent with Appiah's plea for us to treat the personal dimensions of ourselves as "not too tightly scripted," "not too constrained by the demands and expectations of others" (p. 614). But I am more confident than Appiah that such personal life-scripts can and do thrive in the midst of racial expectations. I suppose my goals are more short term, and therefore geared more toward coping with the social reality of race. Racial categories will remain a prominent feature of our social world for the foreseeable future, and will carry expectations for those whom they describe. We can change the expectations, but we cannot jettison racial categories altogether. My hope in applying racial navigation to the gamete market is that the racial sorting of gamete donors might come to mean something different, something less determinate, than it does now. For the racial expectations that parents have for their children affect the identity developments of both children and parents, as well as the broader social norms regarding the role that race should play in family life. The motivations of individual gamete consumers, whatever their race, are morally interesting only insofar as they reinforce a narrow set of racial expectations.

Naomi Zack shares my consequentialist position that no significant moral difference exists between racialism and racism in the construction of family life. There simply is no escaping the fact that human breeding is "a selective practice invented and reinforced within cultures."[40] A white identity is achieved by looking at one's genealogy and finding no black genetic ties. Here, again, the distinction between genetic ties and genetic traits is critical, since many white people exhibit "black" genetic traits. Thus a white person's preference for a white donor as a means to having a white child perpetuates the exclusionary proposition that white families are racially pure. The desire for a white family results in "tightly scripted" identity narratives that limit the life opportunities for everyone.

COPING WITH RACIAL CATEGORIES

Binding racial narratives are of course not unique to the practice of paid gamete donation. The mixture of science with the desire for a racially specific family does, however, exacerbate this general social problem. The new reproductive technologies expose familial racial expectations in a public and systematic way. As Stephen Gould observes, people often turn to science to confirm their racial prejudices, with the effect that these prejudices drive scientific research and the social use of new technologies.[41] The negative consequence is that people will feel more justified in holding race-specific views about family formation.

Zack and Appiah's philosophical arguments against racialism are sound, but their rejection of racial categories is not practical, and therefore not helpful in making today's gamete market more racially just. Philosophers interested in bioethics need to offer some direction on how to cope with racial categories in the here and now instead of turning to utopian visions of a world without race. The most we can do to alleviate the cruelty of racial categories today and in the future is to find ways to maximize human agency under racial constraint in the hopes of breaking the stereotypes that feed personal and social race-based harm. This flexibility in light of racial imposition is the purpose of racial navigation, which aims to mediate between the racial expectations of others and one's own path.

I am not proposing that breaking racial stereotypes through examples of varied racial response is all that is needed to fight the invidious effects of racialism in the gamete market and beyond. But racial navigation is a vital piece of the puzzle. Laws can and should set the parameters for our social conduct, but law cannot (nor should it try to) reach into the realm of human intimacy. That domain is for us to grapple with in our own hearts and minds. We have a moral responsibility to question the racial meanings swirling around in our social ontology because these racial meanings shape and constrict our self-understandings, and by extension, our family plans.

NOTES

I wish to thank those who participated in the Program on Ethics and Public Life 2000 Young Scholar Weekend at Cornell University, where I presented an earlier draft of this article.

1. J. A. Robertson, *Children of Choice: Freedom and the New Reproductive Technologies* (Princeton: Princeton University Press, 1994).

2. In 1995, the National Survey of Family Growth found that 10.5 percent of African American women were infertile, compared to 6.4 percent of white women. Seven percent of Hispanic women were infertile, and "other" women experienced infertility at a rate of 13.6 percent. "Fertility, Family Planning, and Women's Health: New Data From the 1995 National Survey of Family Growth," *U.S. Dept. of Health and Human Services.* High rates of infertility among black women may be linked to "untreated chlamydia and gonorrhea, STDs that can lead to pelvic inflammatory disease; nutritional deficiencies; complications of childbirth and abortion; and environmental and workplace hazards"; D. Roberts, *Killing the Black Body: Race, Reproduction, and the Meaning of Liberty* (New York: Random House, 1977), pp. 251–52.

3. In this article I focus on economic transactions involving sperm and ova, but the race-based selection and "hiring" of surrogate and gestational mothers also raises ethical concerns. As with paid gamete "donation," the reinforcement of racial stereotype is of concern, but issues of economic exploitation that entangle race and class status present additional moral worries.

4. The term "donor" continues to be used to describe individuals who are in fact paid in exchange for their gametes. "Donor" attributes altruistic motives to the sellers of gametes even when financial gain is the only or overriding factor expressed. See R. Mead, "Eggs for Sale," *The New Yorker* (9 August 1999).

5. Doctors must be licensed to perform fertility therapies such as in vitro fertilization and gamete intrafallopian transfer, and clinics must report their success rates to the Centers for Disease Control. The American Society of Reproductive Medicine recommends guidelines and ethical standards, but physicians are not legally obligated to follow their recommendations.

6. N. Zack, *Race and Mixed Race* (Philadelphia: Temple University Press, 1993), p. 40.

7. I develop the concept of racial navigation in the context of the lingering public debate over transracial adoption in the book *The Ethics of Transracial Adoption* (Ithaca: Cornell University Press, forthcoming 2001).

8. In 1988, 12 percent of U.S. women of childbearing age sought professional advice regarding infertility (medical advice, tests, drugs, surgery, or assisted reproductive technologies). This number grew to 15 percent in 1995. See ref. 2, "Fertility, Family Planning, and Women's Health." Sperm counts in U.S. men have decreased annually from 1938 to 1996 at a rate of about 1.5 percent. The National Institutes of Health, "Environmental Health Perspectives," November 1997.

9. http://news.mpr.org/features/199711/20-smiths-fertility/part3/section1.shtml.

10. See ref. 8, "Fertility, Family Planning, and Women's Health."

11. http://news.mpr.org/features/199711/20-smiths-fertility/part3/

12. This cost covers the entire process from the initial consultation to the actual transfer. American Society for Reproductive Medicine, 1995; see website for the American Society for Reproductive Medicine, Patient FAQ.

13. See ref. 2, Roberts, *Killing the Black Body,* p. 251.

14. These other factors include the intergenerational "grand shuffling of the genetic deck," which "gives each gene slightly different properties and is one reason children differ from their parents." N. Wade, "Earliest Divorce Case: X and Y Chromosomes," *New York Times* (29 October 1999).

15. For a historical overview of the "one drop rule" see J. F. Davis, *Who Is Black? One Nation's Definition* (University Park, Penn.: The Pennsylvania State University Press, 1991).

16. For a discussion of extended kin networks in black communities see C. Stack, *All Our Kin: Strategies for Survival in a Black Community* (New York: Harper and Row, 1974).

17. Orlando Patterson contends that African Americans are a diverse group and "very American." "They are a hard-working, disproportionately God-fearing, law-abiding group of people who share the same dreams as their fellow citizens, love and cherish the land of their birth with equal fervor, contribute to its cultural, military, and political glory and global triumph out of all proportion to their numbers, and, to every dispassionate observer, are, in their values, habits, ideals, and ways of living, among the most 'American' of Americans." *The Ordeal of Integration* (New York: Civitas, 1997), p. 171.

18. Dorothy Roberts documents the sterilization abuse of black women that ironically increased with the demise of Jim Crowism. Physicians at state institutions often performed hysterectomies on poor black women without their consent. Roberts also details the recent efforts to require the injection of Norplant and Depo-Provera as a condition to receiving welfare benefits. See ref. 2, Roberts, *Killing the Black Body,* p. 4.

19. Although my Internet search of fertility clinics offering sperm and egg donation was not exhaustive, every site I visited used race as the primary sorting category. Not all clinics make their donor lists available to the public, and some require registration with the clinic as a precondition to viewing donor lists.

20. Available at http://db.otn.com/cryobank/cryo.frn.

21. C. Mills, *Blackness Visible* (Ithaca: Cornell University Press, 1998), p. 44.

22. What I term "structural racism" is similar to what Kwame Ture and Charles Hamilton call "institutional racism." I prefer the former because it denotes a metaphysical as well as physical rendering of racial hierarchy.

23. D. Bell, *Faces at the Bottom of the Well: The Permanence of Racism* (New York: Basic Books, 1992).

24. M. Banton, "The Idiom of Race: A Critique of Presentism," in *Theories of Race and Racism,* ed. L. Back and J. Solomos (New York: Routledge, 2000).

25. T. Derricotte, *The Black Notebooks: An Interior Journey* (New York: W.W. Norton, 1997); see *Village of Arlington Heights v. Metropolitan Hous. Dev. Corp.,* 429 U.S. 252 (1977).

26. See ref. 1, Robertson, *Children of Choice,* p. 20.

27. For example, R. Posner, *Sex and Reason* (Cambridge, Mass.: Harvard University Press, 1992).

28. R.D.G. Kelley, *Yo' Mama's DisFunktional! Fighting the Culture Wars in Urban America* (Boston: Beacon Press, 1997), p. 16.

29. For example, W. Kymlicka's argument in favor of "contexts of choice" in *Liberalism, Community, and Culture* (New York: Oxford University Press, 1989).

30. See ref. 28, Kelley, *Yo' Mama's DisFunktional!,* p. 17.

31. R. Gooding-Williams, "Race, Multiculturalism and Democracy," *Constellations* 5, no. 1 (1998): 23.

32. The strategic value of racial self-awareness is tangible in the domain of antidiscrimination laws and policies that attempt to correct for race-based disadvantage. This critique of colorblindness is found in the body of work that has been dubbed Critical Race Theory. For a thorough overview of the scholar-activist movement see the introduction to K. Crenshaw et al., eds., *Critical Race Theory: The Key Writings That Made the Movement* (New York: New Press, 1996).

33. J. Scales-Trent, *Notes of a White Black Woman* (University Park, Penn.: Pennsylvania State University Press, 1995).

34. See ref. 31, Gooding-Williams, "Race, Multiculturalism and Democracy," p. 23.

35. For example, A. Piper. "Passing for White, Passing for Black," *Transition* 58 (1992): 4–32.

36. For an excellent example of contemporary "whiggers" see the film "Black and White."

37. The issue of parental rights and obligations of genetic and social parents is not a settled matter. Ethical and legal issues become especially controversial when the custody or ownership of embryos is at stake.

38. A. Appiah, *In My Father's House* (New York: Oxford. 1992). p. 13.

39. A. Appiah, "Racial Identity and Racial Identification," in *Theories of Race and Racism,* ed. L. Back and J. Solomos (New York: Routledge, 2000), p. 613.

40. See ref 6, Zack, *Race and Mixed Race,* p. 40.

41. S.J. Gould, *The Mismeasure of Man* (New York: Norton, 1981).

9

How Can You Patent Genes?

Rebecca S. Eisenberg

As rival initiatives in the public and private sectors race to complete the sequence of the human genome (Ross 2000, 98; Smaglik 1999; Collins 2000; Gillis and Weiss 1998; Wade 1998), patent issues have played a prominent role in speculations about the significance of this achievement (Gosselin 2000b; King 2000; Gillis 1999). How much of the genome will be subject to the control of patent holders, and what will this mean for future research and the development of products for the improvement of human health (Gosselin 2000a)? Is a patent system developed to establish rights in mechanical inventions of an earlier era up to the task of resolving competing claims to the genome on behalf of the many sequential innovators who elucidate its sequence and function, with due regard to the interests of the scientific community and the broader public ("Human Genes without Functions" 1993; Ducor 1997; Fields 1997; Eisenberg 1997; Enserink 2000; Merz et al. 1997)?

A deeper question is logically prior to these more fine-grained inquiries: How can you patent DNA sequences? Indeed, over the course of fifteen years of giving talks on the topic of biotechnology patents to widely varying audiences, this has been the question that I am asked most frequently and persistently. Although patent applicants have been seeking and obtaining patent claims on DNA sequences for some 20 years already, many people find this practice troubling and counterintuitive (Eisenberg 1990). One might expect that the Patent and Trademark Office (PTO) and the courts would have resolved this issue many times over as the industry has pursued and litigated patent claims covering biotechnology products; one might also expect that biotechnology patent law would now be entering a relatively mature phase in which fundamental questions have been resolved and the issues that remain to be addressed are incremental and interstitial.[1] Instead, the patent system is struggling to clarify the ground rules for patenting DNA sequences, while years' worth of patent applications accumulate in the PTO. What accounts for how patent law applies to this technology?

A significant part of the problem lies in the shifting landscape of discovery in genetics and genomics research. The patent system, which, inevitably, requires time to resolve even routine matters, has so far focused primarily on the discoveries of the 1980s (Lemley 1995). DNA sequences that were the subject of patent claims in this era typically consisted of cloned genes that enabled the production of proteins through recombinant DNA technology. Patents on the genes and recombinant materials that supply the genetic blueprint for these proteins promised exclusivity in the market for the protein itself, equivalent to the protection that a pharmaceutical firm obtains by patenting a new drug. From this perspective, patents on DNA sequences seemed analogous to patents on new chemical entities. The United States Court of Appeals for the Federal Circuit (Federal Circuit) accordingly turned to prior cases considering patents on chemicals in resolving disputed issues about how patent law should apply to DNA sequences.[2] Whatever the limitations of this analogy, it provided a relatively clear point of departure for analyzing legal issues presented by patent claims on the first generation of biotechnology products that came to market: therapeutic proteins produced through recombinant DNA technology.

As DNA sequence discovery has moved beyond targeted efforts to clone particular genes to large-scale, high-throughput sequencing of entire genomes, new questions have come into view. The DNA sequences identified by high-throughput sequencing look less like new chemical entities than they do like new scientific information. From the perspective of patent claimants, the chemical analogy is of little value as a strategic guide to capturing the value of this information as intellectual property. From the perspective of the PTO and the courts, claims to these discoveries raise unresolved issues on which the chemical analogy sheds little light. The result is profound uncertainty concerning the meaning of the doctrinal tools that the patent system offers for determining what may be patented and for drawing boundaries between the rights of inventors and the rights of the public.

PATENT ELIGIBILITY

A threshold issue that one might expect to have been resolved long ago is whether DNA sequences are the sort of subject matter that the patent system protects. The U.S. patent statute defines patent-eligible subject matter as "any process, machine, manufacture, or composition of matter" (35 U.S.C. sec. 101), language that the U.S. Supreme Court in *Diamond v. Chakrabarty* (447 U.S. 303 [1980]) held to indicate an expansive scope that includes "anything under the sun that is made by man." Although cases have held that "products of nature" may not be patented, this exclusion has not presented an obstacle to obtaining patents claiming DNA sequences in forms that do not occur in nature as new "compositions of matter." On the threshold issue of patent-eligible subject matter, as on other issues, the analogy to chemical patent practice has supplied an answer.

The standard patent lawyer's response to the "products of nature" intuition is to treat it as a technical, claim-drafting problem. From this perspective the prohibition against patenting products of nature only prevents the patenting of DNA sequences in a naturally occurring form that requires no human intervention. One cannot get a patent on a DNA sequence that would be infringed by someone who lives in a state of nature on Walden Pond, whose DNA continues to do the same thing it has done for generations in nature. But one can get a patent on DNA sequences in forms that only exist through the intervention of modern biotechnology.

Patents have thus issued on "isolated and purified" DNA sequences, separate from the chromosomes in which they occur in nature, or on DNA sequences that have been spliced into recombinant vectors or introduced into recombinant cells of a sort that do not exist on Walden Pond.[3] This is consistent with longstanding practice, even prior to the advent of modern biotechnology, of allowing patents to issue on isolated and purified chemical products that exist in nature only in an impure stare, when human intervention has made them available in a new form that meets human purposes.[4] This is not simply a lawyer's trick but a persuasive response to the intuition that patents should only issue for human inventions. It prevents the issuance of patents that take away from the public things that they were previously using (such as the DNA that resides in their cells), while allowing patents to issue on new human manipulations of nature. Those of us who simply use the DNA in our own cells, as our ancestors did for generations, should not and need not worry about patent infringement liability. On the other hand, those of us who get shots of recombinant insulin or erythropoietin can in fairness expect to pay a premium to the inventors who made these technological interventions possible.

MOLECULES VERSUS INFORMATION

The patentability of DNA molecules in forms that involve human intervention appears to be well settled. But recent advances in DNA sequencing present the patent-eligibility issue from a somewhat different angle that the courts have yet to address. DNA sequences are not simply molecules; they are also information. Knowing the DNA sequence for the genome of an organism provides valuable scientific information that can open the door to future discoveries. Can the value of this information be captured through patents? Can information about the natural world, as distinguished from tangible human interventions that make use of that information, be patented?

The traditional statutory categories of patent-eligible subject matter — processes, machines, manufactures, and compositions of matter — seem to be limited to tangible products and processes, as distinguished from information as such. Although many cases have used the word "tangible" in defining the boundaries of patentable subject matter, neither the language of the statute nor judicial decisions elaborating its meaning have explicitly excluded "information" from patent protection. Arguably, such a limitation is implicit in prior judicial decisions stating that the patent system protects practical applications rather than fundamental new insights about the natural world.[5] The exclusion of information itself from patent protection is also at least implicit in the statutory requirement that patent applicants make full disclosures of information about their inventions, with no restrictions upon public access to the disclosure once the patents issue.[6]

Patent claims on DNA sequences as "compositions of matter" give patent owners exclusionary tights over tangible DNA molecules and constructs but do not prevent anyone from perceiving, using, and analyzing information about what the DNA sequence is. Once the patent issues, this information becomes freely available to the world, subject only to the inventor's right to exclude others from making and using the claimed compositions of matter. For patents on genes that encode therapeutic proteins, the value of this exclusionary right over tangible compositions of matter has been sufficiently large (relative to the value of the information that spills over to the public through the patent disclosure) to motivate inventors to file patent applications rather than to keep the sequence secret.

By contrast, in the setting of high-throughput DNA sequencing, the informational value of knowing what the sequence is often exceeds the tangible value of exclusionary rights in DNA molecules and constructs, at least initially. This information base provides a valuable resource for future discovery, only part of which corresponds to those portions of the sequence-encoding proteins. DNA molecules corresponding to the portions of the sequence that encode valuable proteins may ultimately prove valuable as tangible compositions of matter. But it might not be immediately apparent just which parts of the sequence encode which proteins and what, if anything, makes those proteins valuable.

It is not obvious how an inventor might use patents to capture the value of the sequence under these circumstances. It may be difficult to draft claim language that covers the portions of the sequence that prove to have tangible value. An inventor who claims too broadly runs the risk that the claim will be invalid because it covers similar sequences that have already been disclosed in the prior art. An inventor who claims too narrowly runs the risk that the claim will prove easy to evade through minor changes in the molecule.[7] More important, claim language that is directed to tangible molecules fails to capture the informational value of knowing the sequence itself. If this informational value is large relative to the speculative value of tangible molecules corresponding to portions of the sequence, the more sensible strategy may be to sell access to a proprietary database of sequence information. So far, database subscriptions have been the principal source of revenue for most private firms involved in high-throughput DNA sequencing, although the same firms have also filed patent applications.

CLAIMING COMPUTER-READABLE INFORMATION

Another strategy that the PTO and the courts have yet to consider seeks to capture the informational value of DNA sequences through patent claims directed toward sequences stored in a computer-readable medium. An early example of this strategy is the patent application filed by Human Genome Sciences (HGS) on the sequence of the *Haemophilus influenzae* Rd genome.[8] This patent application has not yet issued as a patent anywhere in the world, but it was published eighteen months after its filing date under the terms of article 21(2) of the Patent Cooperation Treaty of 19 June 1970. *Haemophilus influenzae* is a bacterial strain that causes ear and respiratory tract infections in humans. It was the first organism whose genome was fully sequenced, and the fate of the related patent applications may offer a preview of how the patent system will allocate patent rights in future genomic discoveries (Fleischmann et al. 1995). HGS filed a patent application setting forth the complete nucleotide sequence of the genome, identified as "SEQ ID NO. 1."[9] The application concludes with a series of claims representing the invention to which HGS seeks exclusive rights. The first of these claims reads as follows: "Computer readable medium having recorded thereon the nucleotide sequence depicted in SEQ ID NO. 1, a representative fragment thereof or a nucleotide sequence at least 99.9% identical to the nucleotide sequence depicted in SEQ ID NO. 1." It bears emphasizing that this is not yet an issued patent. The foregoing claim language, in effect, is the first item on the wish list of HGS for patent rights associated with the discovery of the *H. influenzae* genome.

This claiming strategy represents a fundamental departure from the previously sanctioned practice of claiming DNA sequences as tangible molecules. By claiming exclusionary rights in the sequence information itself, if stored in a computer-readable medium, HGS seeks patent rights that would be infringed by information storage, retrieval, and analysis rather than simply by making, using or selling tangible molecules.[10] It remains to be seen whether the PTO will issue such a claim, or whether a rejection would stand up on appeal to the Federal Circuit.[11]

EXPANSIVE TREND OF CASE LAW

Recent decisions concerning the patentability of computer-implemented inventions may provide more guidance than prior decisions concerning the patentability of discoveries in the life sciences in predicting whether DNA sequence information stored in a computer-readable medium may be patented. The overall trend of decisions in the Federal Circuit is toward expansive interpretation of the scope of patent-eligible subject matter—even for categories of inventions that prior decisions seemed to exclude from the protection of the patent statute—in order to make the patent system "responsive to the needs of the modern world" *(AT&T Corp. v. Excel Communications, Inc.,* 172 F.3d 1352 [Fed. Cir. 1999]). The most conspicuous recent example of this trend was the 1998 decision in *State Street Bank & Trust v. Signature Financial Group* (149 F.3d 1368 [1998], *cert. denied,* 199 S.Ct. 851 [1999]) upholding the patentability of a computer-implemented accounting system for managing the flow of funds in partnerships of mutual funds that pool their assets. This invention arguably fell within previously apparent judicial limitations that excluded mathematical algorithms and business methods from patent protection.[12] Yet the Federal Circuit minimized the first of these limitations, holding that it excluded from patent protection only "abstract ideas constituting disembodied concepts or truths that are not 'useful'" and repudiated the second, insisting that "[t]he business method exception has never been invoked by this court, or [its predecessor], to deem an invention unpatentable" and that other courts that had appeared to apply the business method exception always had other grounds for arriving at the same decision.

Rather than seeing the language of section 101 of the Patent Act, which permits patents to issue for "any new and useful process, machine, manufacture, or composition of matter," as a significant limitation on the types of advances that might qualify for patent protection, the Federal Circuit characterizes this language as a "seemingly limitless expanse," subject only to three "specifically identified . . . categories of unpatentable subject matter: 'laws of nature, natural phenomena, and abstract ideas"' *(AT&T Corp. v. Excel Com-munications, Inc.,* 172 F.3d 1352 [Fed. Cir. 1999], citing *Diamond v. Diehr,* 450 U.S. 175 [1981]). In this environment it is not obvious why DNA se-quence information stored in a computer-readable medium would be categor-ically excluded from patent protection.

PTO GUIDELINES

Of course, DNA sequence information stored in a computer-readable medium is not the same thing as a computer-implemented business method, and it is certainly possible to define boundaries for the patent system that include the latter but not the former. Indeed, the PTO's Examination Guidelines for Computer-Implemented Inventions exclude data stored in computer-readable medium from patent protection *(Fed. Reg.* 1996, 61:7478). The Guidelines distinguish between "functional descriptive material" (such as "data struc-tures and computer programs which impart functionality when encoded on a computer-readable medium") and "non-functional descriptive material" (such as "music, literary works and a compilation or mere arrangement of data [that] is not structurally and functionally interrelated to the medium but is merely carried by the medium").[13] Although functional descriptive material will generally fall within the statutory categories of patent-eligible subject matter, the Guidelines state that nonfunctional descriptive material will gen-erally not meet the statutory limitations; "Merely claiming non-functional de-scriptive material stored in a computer-readable medium does not make it statutory. Such a result would exalt form over substance."[14] DNA sequence information stored in a computer-readable medium seems to fall squarely within the PTO's definition of "non-functional descriptive material" that is "merely carried by" the computer-readable medium and is not functionally interrelated to it.

If the PTO continues to follow these four-year-old guidelines, it should reject claims to DNA sequence stored in a computer-readable medium. But if a dis-gruntled patent applicant appeals to the Federal Circuit, that court might well re-verse the rejection. The distinction between tangible molecules and intangible information may do little work today in delineating the boundaries of patent el-igibility in the face of recent decisions deemphasizing the importance of physi-cal limitations in establishing the patentability of computer-implemented inven-tions. This shift in emphasis is particularly apparent in *AT&T v. Excel Communications,* in which the court explicitly declined to focus on the "physi-cal limitations inquiry" that had played a central role in distinguishing between unpatentable mathematical algorithms and patentable computer-implemented ` inventions in its prior decisions (172 F.3d 1352 [Fed. Cir. 1999]). Instead, the court asked "whether the mathematical algorithm is applied in a practical manner to produce a useful result" (172 F.3d 1352 [Fed. Cir. 1999]). This approach appears to merge the issue of patent eligibility with the issue of utility, opening the door to patent claims to information so long as it is "useful."

TRADITIONAL PATENT BARGAIN

If the Federal Circuit steps back from the momentum of its recent decisions expanding the boundaries of the patent system, it should not be persuaded by this argument. Patent claims to information stored in a computer-readable medium represent a fundamental departure from the traditional patent bargain. That bargain calls for free disclosure of information to the public at the outset of the patent term in exchange for exclusionary rights in particular tangible applications until the patent expires. Claims that are infringed by mere perception and analysis of the information set forth in the patent disclosure undermine the strong policy preventing patent applicants from restricting access to the disclosure once the patent has issued.[15] The limitation that the information be stored in "a computer-readable medium" offers scant protection for the public interest in free access to the informational content of patent disclosures. Scanning technologies arguably bring paper printouts of DNA sequence information within the scope of the claim language, an interpretation that would make copying the patent document itself an act of infringement. Even if the claim language is more narrowly interpreted to cover only electronic media, numerous websites post the full text of issued patents, including a website maintained by the U.S. Patent and Trademark Office.[16] Any claim that would count these postings as acts of infringement simply proves too much.

That patents on information represent a departure from tradition may not be a sufficient ground to reject them in light of the increasing importance of information products to technological progress. Perhaps the traditional bargain of free disclosure of information in exchange for exclusionary rights in tangible applications doesn't make sense in this new environment. If the value of unprotectible information gained from high-throughput DNA sequencing is large relative to the value of tangible molecules that might be covered by established claiming strategies, patents that do not allow the inventor to capture the value of the information might not do enough to motivate investment in DNA sequencing. This may seem unlikely as an empirical matter, given the substantial investments that are being made in DNA sequencing efforts in both the public and private sectors with no clear precedent for capturing the informational value of this investment through the patent system, but it is at least a logical possibility. A more plausible speculation is that inventors might forgo the patent bargain if they are stuck with the traditional terms of that bargain, choosing instead to exploit their discoveries through restricted access to proprietary DNA sequence databases.

On the other hand, if the terms of the bargain are altered to allow patent holders to capture the informational value of their discoveries, the bargain becomes less attractive to the public. If the issuance of patents does not leave the public free to perceive and analyze the information disclosed in patent specifications, the public might be better off withholding patents and allowing others to derive the same information independently. Withholding patents makes particular sense if the efforts of the patent holder are not necessary to bring the information into the public domain. Much DNA sequence information is freely disclosed in the public domain, both by publicly funded researchers and by private firms. If a discovery is likely to be made and disclosed promptly even without patent incentives, there is little point in enduring the social costs of exclusionary rights.[17]

BRICKS-AND-MORTAR RULES FOR INFORMATION GOODS

There are sound policy reasons to be wary of permitting use of the patent system to capture the value of information itself. The traditional patent bargain ensures that patenting enriches the information base, even as it slows down commercial imitation.[18] This balances the interests of the inventor in earning a return on past research investments against the interest of the larger public in avoiding impediments to future research. If patent claims could prevent the perception and analysis of information, this balance would tilt sharply in favor of patent owners.

More generally, patents are a form of intellectual property right that is particularly ill suited to the protection of information because there are so few safety valves built into the patent system to constrain the rights of patent holders in favor of the competing interests of the public. Unlike copyright, patent law has no fair-use defense that permits socially valuable uses to go forward without a license.[19] Contrary to the understanding of many scientists, patent law has only a "truly narrow" research exemption that offers no protection from infringement liability for research activities that are commercially threatening to the patent holder.[20] Nor is independent creation a defense to patent infringement. Unlike trade-secret law, patent law has no defense for reverse engineering. The most important safety valve built into a patent, apart from its finite term,[21] is the disclosure requirement that permits unlicensed use of information about the invention, as distinguished from the tangible invention itself. But if patents issue that restrict the public from perceiving and analyzing the information, the claim effectively defeats that safety valve.

The foregoing discussion might seem to understate the implications for patents on DNA sequences of a principle that excludes information from patent protection. The DNA molecule itself may be thought of as a tangible storage medium for information about the structure of proteins. Cells read the information stored in DNA molecules to make the proteins that they need to survive in their environments, and they copy that information when they divide and reproduce. If DNA sequence information is not patentable when it is stored in a medium that is readable by computers, how can it nonetheless be patented when stored in a medium that is readable by living cells?

A quick answer is that information stored in a computer-readable medium is directed at the human observers who are the intended beneficiaries of the information spillovers that arise through patent disclosures. It is therefore *human* readable information that must not be patented as such in order to maintain a balance between the exclusionary rights of patent holders and the rights of the public to use the disclosures that are the quid pro quo of those exclusionary rights. But humans can direct queries to DNA sequence information whether it is stored in molecular form or in electronic form. One might, for example, use DNA molecules as probes to detect the presence of a particular DNA sequence in a sample. This sort of molecular query has diagnostic and forensic applications as well as research applications. Researchers seeking to learn more about the functional significance of DNA sequence information are likely to query the information in both computer-readable and molecular form (Ekins and Chu 1999; Sinclair 1999). The distinction between computer-readable and molecular versions of DNA sequence is particularly difficult to maintain in the context of DNA array technology. DNA array technology involves immobilizing thousands of short oligonucleotide molecules on a substrate to detect the presence of particular sequences in a sample using specialized robotics and imaging equipment.[22] In effect, this technology enables people to use computers to perceive information stored in DNA molecules in a sample. When contemporary technology blurs the boundaries between computer-readable and molecular forms of DNA, what logic is there to drawing this distinction in determining the patent rights of DNA sequencers?

Perhaps the best argument for maintaining a distinction between DNA sequence information and DNA molecules at this point in the history of patents for genetic discoveries is consistency with tradition and precedent. Any categorical exclusion of DNA molecules from eligibility for patent protection would contradict the practice of the PTO and the courts for two decades and would undermine the precedent-based expectations of a patent-sensitive industry. On the other hand, allowance of patent claims to DNA sequence information stored in a computer-readable medium would extend patentable subject matter beyond what the PTO and the courts have recognized thus far, departing from a longstanding tradition of free access to the information disclosed in issued patents.

This analysis may seem stubbornly bricks-and-mortar in its focus on tangibility as the touchstone for protection—and therefore out of step with the needs of the modern information economy. If a significant portion of the value of DNA sequencing resides in the information that it yields, rather than in the molecules that correspond to that information, then perhaps we should not assume that those investments will be forthcoming on the basis of an intellectual property system that limits exclusionary rights to tangible things and allows the value of the information itself to spill over to the general public.[23] At some point we may need intellectual property rights that permit the creators of information products to capture the value of the information itself in order to motivate socially valuable investments. But if we have arrived at that point, then we need to look beyond the patent system for a suitable model. The patent system was designed to serve the needs of a bricks-and-mortar world, and it would be foolish to assume that it can meet the changing needs of the information economy simply by expanding the categories of subject matter that are eligible for patent protection.

NOTES

Adapted from *Who Owns Life?*, eds. Magnus, Caplan, and McGee, Prometheus Books 2003. By permission of Prometheus Books.

1. See, e.g., *Amgen v. Chugai Pharmaceutical Co.*, 927 F.2d 1200 (Fed. Cir. 1991) (erythropoietin); *Scripps Clinic & Research Found. v. Genentech*, 927 F.2d 1565 (Fed. Cir. 1991) (Factor VIII:C); *Genentech v. The Welcome Found.*, 29 F.2d 1555 (Fed. Cir. 1994) (tissue plasminogen activator); *Hormone Research Found. v. Genentech* 904 F.2d 1558 (Fed. Cir. 1990) (human growth hormone); *Novo Nordisk v. Genentech*, 777 F.3d 1364 (Fed. Cir. 1996) (human growth hormone); *Genentech v. Eli Lilly & Co.*, 998 F.2d 931 (Fed. Cir. 1993) (human growth hormone); *Bio-Technology General v. Genentech*, 80 F.3d 1553 (Fed. Cir. 1996) (human growth hormone); and *Enzo Biochem v. Calgene*, 188 F.3d 1362 (Fed. Cir. 1999) (Flavr Savr® tomato).

2. See, e.g., *Amgen v. Chugai Pharmaceutical Co.*, 927 F.2d 1200 (Fed. Cir. 1991), *cert. denied sub nom. Genetics Institute v. Amgen*, 502 U.S. 856 (1991). "A gene is a chemical compound, albeit a complex one. . . ."

3. *Amgen, Inc. v. Chugai Pharmaceutical Co.*, 13 U.S.P.Q.2d (BNA) 1737 (D. Mass. 1990). "The invention claimed in the '008 patent is not as plaintiff argues the

DNA sequence encoding human EPO since that is a nonpatentable natural phenomenon 'free to all men and reserved exclusively to none.' . . . Rather, the invention as claimed in claim 2 of the patent is the 'purified and isolated' DNA sequence encoding erythropoietin."

4. See, e.g., *Merck & Co. v. Olin Mathieson Chemical Corp.*, 253 F.2d 156 (4th Cir. 1958), upholding the patentability of purified vitamin B12.

5. See, e.g., *Diamond v. Chakrabarty* ("Einstein could not patent his celebrated law that $E = mc^2$ nor could Newton have patented the law of gravity. Such discoveries are 'manifestations of . . . nature, free to all men and reserved exclusively to none.'"); and *Dickey-John Corp. v. International Tapetronics Corp.*, 710 F.2d 329 (7th Cir. 1983) ("Yet patent law has never been the domain of the abstract—one cannot patent the very discoveries which make the greatest contributions to human knowledge, such as Einstein's discovery of the photoelectric effect, nor has it ever been considered that the lure of commercial reward provided by a patent was needed to encourage such contributions. Patent law's domain has always been the application of the great discoveries of the human intellect to the mundane problems of everyday existence.")

6. 35 U.S.C. secs. 112, 154(aX4). See *In re Argoudelis*, 434 F.2d 1390 (Ct. Customs & Pat. App. 1970).

7. The language of patent claims defines the scope of the patent holder's exclusionary rights. See 35 U.S.C. sec. 112; and *Ex parte Fressola*, 27 U.S.P.Q.2d (BNA) 1608 (Bd. Pat. App. & Interf. 1993). A broad claim is a claim that has few limitations. One might, for example, seek a claim that covers any molecule that includes (or "comprises," in the vernacular of patent law) at least ten consecutive nucleotides from the disclosed sequence. If allowed, such a claim would be very broad in that it would be likely to cover any portion of the sequence that later proves to encode a valuable protein. But the breadth of the claim makes it more likely that it will be held invalid. The claim would be invalid if any previously disclosed DNA sequence included any 10 consecutive nucleotides that were identical to the portion of the sequence disclosed in the patent application. The shorter the portion of the disclosed sequence that is necessary to establish infringement, the broader the claim. But the broader the claim, the easier it is to find "prior art" disclosures that would fall within the scope of the claim, rendering the claim invalid. A narrow claim is a claim that has many limitations. One might, for example, claim the entire disclosed sequence as an isolated molecule. Since every element of the claim must be present in a competitor's product to establish infringement, a competitor who made a DNA molecule that included only a portion of the disclosed sequence corresponding to a particular protein would not be liable.

8. Nucleotide Sequence of the *Haemophilus influenzae* Rd Genome, Fragments Thereof, and Uses Thereof, WQ 961 33276, PCT/US96105320.

9. The sequencing was done at The Institute for Genomic Research (TIGR), a private, nonprofit organization affiliated with Human Genome Sciences (HGS) at the time. Pursuant to an agreement between TIGR and HGS, patent rights in the *H. influenzae* genome were assigned to HGS.

10. The meaning of this term could be quite broad.

11. An applicant whose claims have been rejected by a PTO examiner twice may appeal to the Board of Patent Appeals and Interferences (35 U.S.C. sec. 134), and an

applicant who is dissatisfied with the decision of the Board of Patent Appeals and Interferences may appeal to the United States Court of Appeals for the Federal Circuit (35 U.S.C. sec. 141).

12. See, e.g., *Gottschalk v. Benson,* 409 U.S. 63 (1972); *Parker v. Flook,* 437 U.S. 584 (1978); and *Hotel Security Checking Co. v. Lorraine Co.* 160 F. 467 (2d Cir. 1908).

13. The focus on functional relationship between data and substrate echoes language from *In re Lowry,* 32 F.3d 1579 (Fed. Cir. 1994), in which the Federal Circuit upheld the patentability of a data structure for storing, using, and managing data in a computer memory. In that case, the Board of Patent Appeals had reversed the examiner's rejection of the claims under 35 U.S.C. sec. 101 as claiming nonstatutory subject matter, and the issue of patentable subject matter was therefore not properly before the court on appeal. Nonetheless, in its analysis of the remaining issues of patentability under 35 U.S.C. secs. 102 and 103, the court drew a distinction between claiming information content and claiming a functional structure for managing information: "Contrary to the PTO's assertion, Lowry does not claim merely the information content of a memory. Lowry's data structures, while including data resident in a database, depend only functionally on information content. While the information content affects the exact sequence of bits stored in accordance with Lowry's data structures, the claims require specific electronic structural elements which impart a physical organization on the information stored in memory. Lowry's invention manages information. As Lowry notes, the data structures provide increased computing efficiency" (1583).

14. This qualification in the Guidelines responds to a rhetorical question posed by Judge Archer in his dissenting opinion from the en banc decision of the Federal Circuit in *In re Alappat,* 33 F.3d 1526 (Fed. Cir. 1996). In that case a majority of the court upheld the patentability of a claim to a computer-implemented mechanism for improving the quality of a picture in an oscilloscope. Judge Archer cautioned against the potential implications of allowing patent claims on mathematical algorithms stored in a computer-readable medium in his dissenting opinion asking rhetorically whether a piece of music recorded on a compact disc or player piano roll would be patentable: "Through the expedient of putting his music on known structure can a composer now claim as his invention the structure of a compact disc or player piano roll containing the melody he discovered and obtain a patent therefore? The answer must be no. The composer admittedly has invented or discovered nothing but music. The discovery of music does not become patentable subject matter simply because there is an arbitrary claim to some structure" (33 F.3d 1554).

15. Id at 1360.

16. See the concurring opinion in *In re Argoudelis,* 434 F.2d 1390, 1394–96 (C.C.P.A. 1970); and *Feldman v. Aunstrup,* 517 F.2d 1351, 1355 (C.C.P.A. 1975), *cert. denied,* 424 U.S. 912 (1976).

17. See http://www.uspco.gov/web/menu/pacs.html.

18. Normally the nonobviousness standard set forth at 35 U.S.C. sec. 103 prevents the issuance of parents on inventions that are highly likely to be made independently by another inventor by excluding from patent protection inventions that would have

been "obvious" to persons of ordinary skill in the field of the invention given the state of the art. This standard fails to serve this important function in the context of DNA sequencing because of decisions of the Federal Circuit upholding the patentability of newly identified DNA sequences discovered through routine work, so long as the prior art did not permit prediction of the structure of the DNA molecule. See *In re Bell,* 991 F.2d 781 (Fed. Cir. 1993); and *In re Deuel,* 51 F.3d 1552 (Fed. Cir. 1995).

19. See *Chisum on Patents,* vol. 3, sec. 7.01: "Full disclosure of the invention and the manner of making and using it on issuance of the patent immediately increases the storehouse of public information available for further research and innovation and as-sures that the invention will be freely available to all once the statutory period of mo-nopoly expires."

20. The quoted words are from the opinion of the Court of Appeals for the Federal Circuit in *Roche Prods., Inc. v. Bolar Pharmaceuticals, Inc.,* 733 F.2d 858 (Fed. Cir.), *cert. denied,* 469 U.S. 856 (1984). For a fuller discussion of the research exemption, see Eisenberg (1989).

21. The rule for determining the expiration date of a U.S. patent was changed in 1995 by the *Uruguay Round Amendments Act,* U.S. Public Law 103-465 (H.R. 5110). Prior to passage of the *Act,* U.S. patents expired 17 years after the date that they were issued, regardless of their application filing dates. The new rule, applicable to U.S. patents issued on the basis of patent applications filed after 8 June 1995, provides for expiration 20 years after their filing dates. 35 U.S.C. sec. 15r.

22. After sequencing DNA, researchers might analyze the sequence in computer-readable form to identify similarities to known sequences and then analyze the sequence in cell-readable form to observe the functional significance of different por-tions of the sequence in a living cell or organism. They might, for example, use DNA molecules as probes to determine when and where an organism expresses a particular portion of its DNA sequence, or they might induce a cell to express a particular DNA sequence in order to learn more about the protein that it encodes, or they might inter-rupt expression of a DNA sequence in an organism and observe the consequences in order to learn more about the functions of the corresponding protein. This sort of in-teraction between analysis of electronic information and observation of how cells use the information characterizes what in recent years has become known as "functional genomics" research. See Hieter and Boguski (1997); and Fields (1997).

23. The classic argument for intellectual property is that exclusionary rights are nec-essary to motivate investments in the creation of goods that are costly to make initially, but cheap and easy to copy once someone else has made the initial investment. As grow-ing volumes of information become freely available on the Internet, this argument seems to be overlooking significant incentives to create and disseminate information outside the intellectual property system. See generally Shapiro and Varian (1999).

REFERENCES

Collins, F. S. 2000. The sequence of the human genome: Coming a lot sooner than you think. Available from: http://www.nhgri.nih.gov/NEWS.

Ducor, P. 1997. Recombinant products and nonobviousness: A typology. *Computer and High Technology Law Journal* 13:1.

Eisenberg, R. S. 1989. Patents and the progress of science: Exclusive rights and experimental use. *University of Chicago Law Review* 56:1017.

———. 1990. Patenting the human genome. *Emory Law Journal* 39:721.

———. 1997. Structure and function in gene patenting. *Nature Genetics* 15:125.

Ekins, R., and F. W. Chu. 1999. Microarrays: Their origins and applications. *Trends in Biotechnology* 17:217.

Enserink, M. 2000. Patent Office may raise the bar on gene claims. *Science* 287:1196.

Fields, S. 1997. The future is function. *Nature Genetics* 15:325.

Fleischmann, R. D., et al. 1995. Whole-genome random sequencing and assembly of *Haemophilus influenzae* Rd. *Science* 269:496–512.

Gillis, J. 1999. Md. gene researcher draws fire on filings; Venter defends patent requests. *The Washington Post,* 26 October, E01.

Gillis, J., and R. Weiss. 1998. Private firm aims to beat government to gene map. *The Washington Post,* 12 May.

Gosselin, P. G. 2000a. Patent Office now at heart of gene debate. *L.A. Times,* 7 February.

———. 2000b. Clinton urges public access to genetic code. *L.A. Times,* 11 February.

Hieter, P., and M. Boguski. 1997. Functional genomics: It's all how you read it. *Science* 278:601.

Human genes without functions: Biotechnology tests the patent utility standard. 1993. *Suffolk University Law Review* 27:1631.

King, R. T., Jr. 2000. Code green: Gene quest will bring glory to some; Incyte will stick with cash. *Wall Street Journal,* 10 February, A1.

Lemley, M. A. 1995. An empirical study of the twenty-year patent term. *AIPLA Quarterly Journal* 22:369.

Merz, J. F., et al. 1997. Disease gene patenting is a bad innovation. *Molecular Diagnosis* 2:299–304.

Ross, P. E. 2000. The making of a gene machine. *Forbes,* 21 February.

Shapiro, C., and H. Varian. 1998. *Information rules: A strategic guide to the network economy.* Cambridge: Harvard Business School Press.

Sinclair, B. 1999. Everything's great when it sits on a chip—A bright future for DNA arrays. *The Scientist* 13(24 May): 18.

Smaglik, P. 1999. A billion base pairs, times two. *The Scientist,* 6 December, 13.

Wade, N. 1998. Scientist's plan: Map all DNA within 3 years. *The New York Times,* 10 May, A1.

10

Monitoring Stem Cell Research

The President's Council on Bioethics
Chapter II: Current Federal Law and Policy

Any overview of the state of human stem cell research under the current federal funding policy must begin with a thorough understanding of that policy. This is not as simple as it may sound. From the moment of its first announcement, on August 9, 2001, the policy has been misunderstood (and at times misrepresented) by some among both its detractors and its advocates. Its moral foundation, its political context, its practical implications, and the most basic facts regarding the policy's implementation have all been subjects of heated dispute and profound confusion. Whether one agrees with the policy or not, it is important to understand it as it was propounded, accurately and in its own terms, in the light also of the historical and political contexts in which it was put forward.

This chapter attempts to place the policy in its proper context: to articulate its moral, legal, and political underpinnings as put forward by its authors and advocates); to offer an overview of its implementation thus far; and to begin to describe its ramifications for researchers and for medicine. By articulating the policy in its own terms, we intend neither to endorse it nor to find fault with it.[i] Indeed, in the next chapter we present an overview of arguments on all sides of the question. Here we mean only to clarify, as far as we are able, the original meaning and purpose of the policy, so as to be better able to monitor its impact.

I. A BRIEF HISTORY OF THE EMBRYO RESEARCH FUNDING DEBATE

The federal government makes significant public resources available to biomedical researchers each year—over $20 billion in fiscal year 2003 alone—in the form of research grants offered largely through the National Institutes of Health (NIH). This level of public expenditure reflects the great esteem in which Americans hold the biomedical enterprise and the value we place on the development of treatments and cures for those who are suffering. But such support is not offered indiscriminately. Researchers who accept federal funds must abide by ethically based rules and regulations governing, among other things, the use of human subjects in research. And some policymakers and citizens have always insisted that taxpayer dollars not be put toward specific sorts of research that violate the moral convictions and sensibilities of some portion of the American public. This has meant that controversies surrounding the morality of some forms of scientific research have at times given rise to disputes over federal funding policy. Among the most prominent examples has been the three-decade-long public and political debate about whether taxpayer funds should be used to support research that involves creating or destroying human embryos or making use of destroyed embryos and fetuses—practices that touch directly on the much-disputed questions of the moral status and proper treatment of nascent human life.

In the immediate aftermath of the Supreme Court's 1973 Roe v. Wade decision legalizing abortion nationwide, some Americans, including some Members of Congress, became concerned about the potential use of aborted fetuses (or embryos) in scientific research. In response to these concerns, the Department of Health, Education and Welfare (DHEW, the precursor to today's Department of Health and Human Services) initiated a moratorium on any potential DHEW sponsorship or funding of research using human fetuses or living embryos. In 1974, Congress codified the policy in law, initiating what it termed a temporary moratorium on federal funding for clinical research using "a living human fetus, before or after the induced abortion of such fetus, unless such research is done for the purpose of assuring the survival of such fetus."[1] Concurrently with that moratorium (and also addressing concerns not directly related to embryo and fetal research), Congress established a National Commission for the Protection of Human Subjects of Biomedical and Behavioral Research. Among its other tasks, Congress explicitly assigned the Commission responsibility for offering guidelines for human fetal and embryo research, so that standards for funding might be established and the blanket moratorium might be lifted. The statutory moratorium was lifted once the Commission issued its report in 1975.[2]

In that report, the Commission called for the establishment of a national Ethics Advisory Board within DHEW to propose standards and research protocols for potential federal funding of research using human embryos and to consider particular applications for funding. In doing so, the Commission looked ahead to the possible uses of in vitro embryos, since the first success-ful in vitro fertilization (IVF) of human egg by human sperm had been ac-complished in 1969.[ii] The Department adopted the recommendation in 1975, established an Ethics Advisory Board, and put in place regulations requiring that the Board provide advice about the ethical acceptability of IVF research proposals. The Board first took up the issue of research on in vitro embryos in full in the late 1970s and issued its report in 1979.[3]

By that time, human IVF techniques had been developed (first in Britain) to the point of producing a live-born child (born in 1978). These techniques, and their implications for human embryo research, raised unique prospects and concerns that were distinct from some of those in-volved in human fetal research. As a consequence, starting in the late 1970s, funding of embryo research and funding of fetal research came to be treated as mostly distinct and separate issues. The Ethics Advisory Board concluded that research involving embryos and IVF techniques was "ethically defensible but still legitimately controverted." Provided that re-search did not take place on embryos beyond fourteen days of development and that all gamete donors were married couples, the Board argued, such work was "acceptable from an ethical standpoint," but the Board decided that it "should not advise the Department on the level of Federal support, if any," such work should receive.[4]

This left the decision in the hands of the DHEW, which decided at that stage not to offer funding for human embryo studies. The Ethics Advisory Board's charter expired in 1980, and no renewal or replacement was put for-ward, creating a peculiar situation in which the regulations requiring the Ethics Advisory Board to review proposals for funding remained in effect, but the Board no longer existed to consider such requests. Funding was therefore rendered impossible in practice. Because the Ethics Advisory Board was never replaced, a de facto ban on funding remained in place through the 1980s.

In 1993, Congress enacted the NIH Revitalization Act, a provision of which rescinded the requirement that research protocols be approved by the non-existent Ethics Advisory Board.[5] This change opened the way in princi-ple to the possibility of NIH funding for human embryo research using IVF embryos. The following year, the NIH convened a Human Embryo Research Panel to consider the issues surrounding such research and to propose guide-lines for potential funding applications. The panel recommended that some areas of human embryo research be deemed eligible for federal funding within a framework of recognized ethical safeguards. It further concluded that the creation of human embryos with the explicit intention of using them only for research purposes should be supported under some circumstances.[6]

President Clinton overruled the panel on the latter point, ordering that embryo creation for research not be funded, but he accepted the panel's other recommendations and permitted the NIH to consider applications for funding of research using embryos left over from IVF procedures.[7]

Congress, however, did not endorse this course of action. In 1995, before any funding proposal had ever been approved by the NIH, Congress attached language to the 1996 Departments of Labor, Health and Human Services, and Education, and Related Agencies Appropriations Act (the budget bill that funds DHHS and the NIH) prohibiting the use of any federal funds for research that destroys or seriously endangers human embryos, or creates them for research purposes.

This provision, known as the "Dickey Amendment" (after its original author, former Representative Jay Dickey of Arkansas), has been attached to the Health and Human Services appropriations bill each year since 1996. Everything about the subsequent debate over federal funding of embryonic stem cell research must be understood in the context of this legal restriction. The provision reads as follows:

> None of the funds made available in this Act may be used for—
>
> (1) the creation of a human embryo or embryos for research purposes; or
>
> (2) research in which a human embryo or embryos are destroyed, discarded, or knowingly subjected to risk of injury or death greater than that allowed for research on fetuses in utero under 45 CFR 46.204 and 46.207, and subsection 498(b) of the Public Health Service Act (42 U.S.C. 289g(b)).[iii]
>
> (3) For purposes of this section, the term 'human embryo or embryos' includes any organism, not protected as a human subject under 45 CFR 46 as of the date of the enactment of the governing appropriations act, that is derived by fertilization, parthenogenesis, cloning, or any other means from one or more human gametes or human diploid cells.[8]

This law effectively prohibits the use of federal funds to support any research that destroys human embryos or puts them at serious risk of destruction. It does not, however, prohibit the conduct of such research using private funding. Thus, it addresses itself not to what may or may not be lawfully done, but only to what may or may not be supported by taxpayer dollars. At the federal level, research that involves the destruction of embryos is neither prohibited nor supported and encouraged.

The Dickey Amendment was originally enacted before the isolation of human embryonic stem cells, first reported in 1998 by researchers at the University of Wisconsin, whose work was supported only by private funds (largely from the Geron Corporation and the University of Wisconsin Alumni Research Foundation). The discovery of these cells and their unique and potentially quite promising properties aroused great excitement both within and beyond the scientific community. It led some people to question the policy of withholding federal funds from human embryo research. Most Members of Congress, however, did not change their position, and the Dickey Amendment has been reenacted every year since. For many of its supporters, the amendment expresses their ethical conviction that nascent human life ought to be protected against exploitation and destruction for scientific research, however promising that research might be, and that at the very least such destruction should not be supported or encouraged by taxpayer dollars.

On its face, the Dickey Amendment would seem to close the question of federal funding of human embryonic stem cell research, since obtaining stem cells for such research relies upon the destruction of human embryos. But in 1999, the General Counsel of the Department of Health and Human Services argued that the wording of the law might permit an interpretation under which human embryonic stem cell research could be funded. If embryos were first destroyed by researchers supported by private funding, then subsequent research employing the derived embryonic stem cells, now propagated in tissue culture, might be considered eligible for federal funding. Although such research would presuppose and follow the prior destruction of human embryos, it would not itself involve that destruction. Thus, the Department's lawyers suggested, the legal requirement not to fund research "in which" embryos were destroyed would still technically be obeyed.[9]

This has generally been taken to be a legally valid interpretation of the specific language of the statute, and indeed the subsequent policies of both the Clinton and Bush administrations have relied upon it in different ways. But some critics of the 1999 legal opinion argued that, though it might stay within the letter of the law, the proposed approach would contradict both the spirit of the law and the principle that underlies it.[10] It would use public funds to encourage and reward the destruction of human embryos by promising funding for research that immediately follows and results from that destruction—thereby offering a financial incentive to engage in such destruction in the future. By so doing, these critics argued, it would at least implicitly state, in the name of the American people, that research that destroys human embryos ought to be encouraged in the cause of medical advance. Supporters of the Clinton administration's proposed approach, however, argued that promoting such research—especially given its therapeutic potential—was indeed an appropriate government function, and that the policy proposed by DHHS was neither illegal nor improper, given the text of the statute and provided that the routine standards of research ethics (including informed consent and a prohibition on financial inducements) were met.[11]

The Clinton administration adopted this course of action and drew up specific guidelines to enact it.[iv] But the guidelines, completed just before the end of the Clinton administration, never had a chance to be put into practice, and no funding was ever provided. Upon entering office in 2001, the Bush administration decided to take another look at the options regarding human embryonic stem cell research policy and therefore put the new regulations on hold, pending review.

In conducting its review, the Bush administration stated that it sought a way to allow some potentially valuable research to proceed while upholding the spirit (and not just the letter) of the Dickey Amendment, a spirit that the President himself has advocated.[12] The expressed hope was that the government, while continuing to withhold taxpayer support or encouragement for the destruction of human embryos, might find a way to draw some moral good from stem cell lines that had already been produced through such destruction—given that this deed, even if immoral, could not now be undone. This is the ethical-legal logic of the present stem cell funding policy: it seeks those benefits of embryonic stem cell research that might be attainable without encouraging or contributing to any future destruction of human embryos.

II. THE PRESENT POLICY

The current policy on federal government funding of human embryonic stem cell research, then, must be understood in terms of the constraints of the Dickey Amendment and in terms of the logic of the moral and political aims that underlie that amendment.

At the time of the policy's announcement, a number of embryonic stem cell lines had already been derived and were in various stages of growth and characterization. The embryos from which they were derived had therefore already been destroyed and could no longer develop further. As President Bush put it, "the life and death decision had already been made."[13]

The administration's policy made it possible to use taxpayer funding for research conducted on those preexisting lines, but it refused in advance to support research on any lines created after the date of the announcement. In addition, to be eligible for funding, those preexisting lines would have had to have been derived from excess embryos created solely for reproductive purposes, made available with the informed consent of the donors, and without any financial inducements to the donors—standard research-ethics conditions that had been attached to the previous administration's short-lived funding guidelines, as well as to earlier attempts to formulate rules for federal funding of human embryo research. The policy denies federal funding not only for research conducted on stem cell lines derived from embryos destroyed after August 9, 2001 (or that fail to meet the above criteria), but also (as the proposed Clinton-era policy would have) for the creation of any human embryos for research purposes and for the cloning of human embryos for any purpose.[v]

The moral, legal, and political grounds of this policy have been hotly contested from the moment of its announcement. Debates have continued regarding its aims, its character, its implementation, and its underlying principles, as well as the significance of federal funding in this area of research. For example, many scientists, physicians, and patient advocacy groups contend that the policy is too restrictive and thwarts the growth of a critical area of research. On the other side, some opponents of embryo research believe the policy is too liberal and legitimates and rewards (after the fact) the destruction of nascent human life. Some ethicists argue that there is a moral imperative to remove all restrictions upon potentially life-saving research; other ethicists argue that there is a moral imperative to protect the lives of human beings in their earliest and most vulnerable stages. These and similar arguments are reviewed in the next chapter. But before one can enter into these debates, it is essential first to understand the relevant elements of the policy itself as clearly and distinctly as possible.

III. MORAL FOUNDATION OF THE POLICY

In articulating its proposed funding policy in 1999 and 2000, the Clinton administration expressed a firm determination that funded research could use only those human embryos that had been left over from IVF procedures aimed at reproduction and that had been donated in accordance with the standards of informed consent and in circumstances free of financial inducements. Provided that these crucial conditions were met, the administration argued that the potential benefits of stem cell research were so great that publicly funded research should go forward. In August of 2000, reflecting on the guidelines put forward by his administration, President Clinton remarked,

> Human embryo research [as approved for funding by the NIH guidelines] deals only with those embryos that were, in effect, collected for in-vitro fertilization that never will be used for that. So I think that the protections are there; the most rigorous scientific standards have been met. But if you just—just in the last couple of weeks we've had story after story after story of the potential of stem cell research to deal with these health challenges. And I think we cannot walk away from the potential to save lives and improve lives, to help people literally to get up and walk, to do all kinds of things we could never have imagined, as long as we meet rigorous ethical standards.[14]

Given the promise of embryonic stem cell research, the existence of many embryos frozen in IVF clinics and unlikely ever to be transferred and brought to term, and the willingness of some IVF patients to donate such embryos for research, the Clinton administration reasoned that research using cell lines derived from these embryos could ethically be supported by federal funds. That position implies, of course, that the destruction of embryos is not inherently or necessarily unethical, or so disconcerting as to be denied any federal support. The Clinton-era NIH Embryo Research Panel put succinctly one form of this view in stating that "the preimplantation human embryo warrants serious moral consideration as a developing form of human life, but it does not have the same moral status as infants and children."[15] If there is sufficient promise or reason to support research, the claim of a human embryo to "serious moral consideration" (or, as others, including some of us, have put it, to "special respect"[16]) could be outweighed by other moral aims or principles.

This (at least implicit) understanding of the moral status of human embryos might be seen to have put the Clinton administration at odds with the principle animating the operative law on this subject (the Dickey Amendment). But given its responsibility to carry out the laws as they are enacted, the administration sought a way to advance research within the limitations set by the statute. Its approach to the funding of embryonic stem cell research, therefore, seems to have sought an answer to this question: How can embryonic stem cell research, conducted in accordance with standards of informed consent and free donation, be maximally aided within the limits of the law? The NIH guidelines published in 2000 represent the answer the Clinton administration found: funding research on present and future embryonic stem cell lines, so long as the embryo destruction itself is done with private funds.

The Bush administration appears to have been motivated by a somewhat different question, arising from what seems to be a different view of the morality of research that destroys human embryos. President Bush put the matter this way, in discussing his newly announced policy in August of 2001:

> Stem cell research is still at an early, uncertain stage, but the hope it offers is amazing: infinitely adaptable human cells to replace damaged or defective tissue and treat a wide variety of diseases. Yet the ethics of medicine are not infinitely adaptable. There is at least one bright line: We do not end some lives for the medical benefit of others. For me, this is a matter of conviction: a belief that life, including early life, is biologically human, genetically distinct and valuable.[vi,17]

While expressing a desire to advance medical research, this argument describes a line that such research should not cross, and therefore past which funding should not be offered. That line, in this context, is the destruction of a human embryo for research purposes. The Bush administration thus appears to share the view that underlies both the word and spirit of the Dickey Amendment. In its approach to the stem cell issue it has sought to answer a question that differs, subtly but significantly, from that formulated by the previous administration: *How can embryonic stem cell research, conducted in accordance with basic research ethics, be maximally aided within the bounds of the principle that nascent human life should not be destroyed for research?*

In seeking to answer that question, the Bush administration (like the Clinton administration) had to take account of the existing situation and—as always in such instances—to mix prudential demands and opportunities with an effort at principled judgment. Given the existence of some human embryonic stem cell lines, derived from human embryos that had already been destroyed, the administration determined that it might not simply have to choose between funding research that relies on the ongoing destruction of embryos (and therefore tacitly supporting and encouraging such destruction by paying for the work that immediately follows it) and funding no human embryonic stem cell research at all. The decision regarding the funding of research on already-derived human embryonic stem cells came down to this question: *Can the government support some human embryonic stem cell research without encouraging future embryo destruction?*

The present funding policy is therefore not an attempt to answer the question of how the government might best advance embryonic stem cell research while conforming to the law on the subject. Rather, it is an attempt to answer the question of how the government might avoid encouraging the (presumptively) unethical act of embryo destruction and still advance the worthy cause of medical research. Whether or not one agrees with the premises defining the question, and whether or not one accepts the logic of the answer, any assessment of the policy must recognize this starting point.

From the very beginning, the policy has been described—even by many of its supporters and defenders—as occupying a kind of middle-ground position in the debate over the morality of embryo research. It has been termed a "Solomonic compromise." But while it may be a prudential compromise on the question of funding, it has been argued that the policy—as articulated by its authors—does not seem to be intended as a compromise on the question of the moral status of human embryos or the moral standing of the act of embryo destruction. In this sense, it appears to be not a political "splitting of the difference" but an effort at a principled solution.[18]

To some extent, the effort reflects a traditional approach in moral philosophy to an ancient and vexing question: Can one benefit from the results of (what one believes to be) a past immoral act without becoming complicit in that act?[vii] The moralists' approach suggests that one may make use of such benefits if (and only if) three crucial conditions are met: (1) Non-cooperation: one does not cooperate or actively involve oneself in the commission of the act; (2) Non-abetting: one does nothing to abet or encourage the repetition of the act, for instance by providing incentives or rewards to those who would perform it in the future; and (3) Reaffirmation of the principle: in accepting the benefit, one re-enunciates and reaffirms the principle violated by the original deed in question.

As a plan for redeeming some good from embryo destruction that has already taken place, while not encouraging embryo destruction in the future, the administration's policy appears at least to seek to address each of these three conditions: (1) No federal funds have been or, by this policy, would be used in the destruction of human embryos for research. (2) By restricting research funding exclusively to embryonic stem cell lines derived before the policy went into effect, the policy deliberately refuses to offer present or future financial or other incentives to anyone who might subsequently destroy additional embryos for research; this is the moral logic behind a central feature of the policy, the cut-off date for funding eligibility (though some argue that by failing to call for an end to privately funded research the policy does not altogether avoid complicity). And (3) the President, in his speech of August 9, 2001, and since (as in the passage quoted above and elsewhere), has reaffirmed the moral principle that underlies his policy and the law on the subject: that nascent human life should not be destroyed for research, even if good might come of it. The policy as a whole draws attention to that principle by drawing a sharp line beyond which funding will not be made available.

Of course, since these terms from the parlance of moral philosophy were not those explicitly employed by the policy's authors, they can go only so far in helping us to understand the policy's foundation. As in any public policy decision, prudence is here mixed with principle, in the hope that the two might reinforce (rather than undermine) each other, and a variety of moral aims are brought together. The desire to afford some aid to a potentially promising field of research moderates what might otherwise have been an at least symbolically stauncher stance against embryo destruction: no public funding whatsoever, even for work on stem cell lines obtained from embryos destroyed in the past. Moreover, the desire to show regard for established principles and standards of ethical research leads to an insistence that, to be approved, stem cell lines must have been drawn from embryos produced for reproduction and obtained with consent and without financial inducements. In these ways, the policy gives some due to competing moral and prudential demands. But the policy's central feature—the announcement date separating eligible from ineligible stem cell lines—holds firm to the principle that *public funds* should not be used to encourage or support the destruction of embryos *in the future*.

It is perhaps worth pointing out that one's attitude regarding the best federal funding policy is not simply determined by one's view regarding the moral standing of human embryos, and that even persons who hold the same view of the moral standing of human embryos may not all agree on the best policy. For example, support for the current policy does not necessarily require a belief that human embryos are persons with full moral standing; and conversely, those who believe that human embryos are persons do not necessarily support the policy. One might believe, for instance, that an embryo is a mystery, not clearly "one of us" but unambiguously a life-in-process, and thus conclude that we should err on the side of restraint (non-destruction) when moral certainty is impossible. Or, one might believe that embryos are not simply persons but are nonetheless either worthy of protection from harm or at least worthy of more respect than ordinary human tissues or animals, and that it would be wrong to begin a massive public project of embryo research that offends the deeply-held beliefs of many citizens. Meanwhile, an individual who believes that human embryos have the same moral standing as children or adults may be deeply unsatisfied with the present policy, since merely denying federal encouragement for future embryo destruction while taking no action to prevent privately-funded stem cell research that destroys embryos may be an insufficient response to the ongoing destruction of nascent human life.

For some of its supporters, the policy goes as far as it seems possible to go within the bounds of the spirit and aims of the law—that the government should not encourage or support the destruction of nascent human life for research. Yet at the same time, it goes farther than the federal government has gone before in the direction of actually funding research involving human embryos, since no public funds had ever before been spent on such research. To go further—say, by funding research on the currently ineligible lines derived after August 9, 2001—would not extend the logic of the policy or of the law, but rather contradict them both: it would be a difference not of degree but of principle. By implying that research using embryos destroyed in the future might one day be supported with public funding, such a policy shift would at least implicitly encourage the very act (embryo destruction) that the current policy aims not to encourage. Of course, such a change might well be in order, but the case for it must address itself to the moral argument and its principles, and not only to the state of research and its progress or promise.

Rather than focus on this principled aspect of the policy, the public debate has tended to concentrate on the precise balance of benefits and harms resulting from the combination of the administration's policy and the state of the relevant science. It has focused on whether there are "enough" cell lines or on whether the science is advancing as quickly as it could. And it has proceeded as though the administration's aim was simply to maximize progress in embryonic stem cell research without transgressing the limits of the letter of the law.

Had the decision been based on that aim alone, then claims or evidence of slowed progress alone might, in themselves, constitute an effective argument against it on its own terms (on the ground that the law technically permits federal funding of research on cells derived from embryos whose destruction was underwritten by private funding). But if one accepts the premise that the decision was grounded also in a discernible (albeit highly controversial) moral aim, one cannot show that the policy is wrong merely by pointing to the potential benefits of stem cell research or the potential harm to science caused by restrictions in federal funding. The present policy aims to support stem cell research while insisting that federal funds not be used to support or encourage the future destruction of human embryos. To argue with that policy on its own terms, therefore, one would need to argue with its view of the significance of that aim. Concretely, this means arguing with its ethical position regarding the destruction of nascent human life and with its ethical-political position regarding the significance of government funding of a contested activity.

This latter point—regarding the meaning of government funding—is much neglected in the current debates and deserves further clarification. That will require delving into the important distinction between government permission (that is, an absence of prohibitions) of an activity and government support for an activity. This ethical-political distinction lies at the heart of the stem cell debate.

IV. THE SIGNIFICANCE OF FEDERAL FUNDING

The national debate over human embryonic stem cell research often raises the most fundamental questions about the moral status of human embryos and the legitimacy of research that destroys such embryos. Yet, looking over this debate, it is easy to forget that the question at issue is not whether research using embryos should be allowed, but rather whether it should be financed with the federal taxpayer's dollars.

The difference between *prohibiting* embryo research and *refraining from funding* it has often been blurred by both sides to the debate. Ignored in the battles over embryo research itself, the ethical-political question regarding funding is rarely taken up in full.

That question arises because modern governments do more than legislate and enforce prohibitions and limits. In the age of the welfare state, the government, besides being an enforcer of laws and a keeper of order, is also a major provider of resources. Political questions today, therefore, reach beyond what ought and ought not be allowed. They include questions of what ought and ought not be encouraged, supported, and made possible by taxpayer funding. The decision to fund an activity is more than an offer of resources. It is also a declaration of official national support and endorsement, a positive assertion that the activity in question is deemed by the nation as a whole, through its government, to be good and worthy. When something is done with public funding, it is done, so to speak, in the name of the country, with its blessing and encouragement.

To offer such encouragement and support is therefore no small matter. The federal government is not required to provide such material support, even for activities protected by the Constitution, let alone for those permitted but not guaranteed.[19] The affording of most federal funding is entirely optional, and the choice to make such an offer is therefore laden with moral and political meaning, well beyond its material importance. In the age of government funding, the political system is sometimes called upon to decide not only the lowest standards of conduct, but also the highest standards of legitimacy and importance. When the nation decides an activity is worth its public money, it declares that the activity is valued, desired, and favored.

The United States has long held the scientific enterprise in such high regard. Since the middle of the twentieth century, the federal government, with the strong support of the American people, has funded scientific research to the tune of many hundreds of billions of dollars. The American taxpayer is by far the greatest benefactor of science in the world, and the American public greatly values the contributions of science to human knowledge, human health, and human happiness. And we Americans have overwhelmingly been boosters of medical science and medical progress, deeming them worthy of support for moral as well as material reasons.

But this enthusiasm for medical science is not without its limits. As already noted, we attach restrictions to federally funded research, for instance to protect human subjects. In fact at times we even use funding to *place* restrictions on research that might otherwise not be constrained. Indeed, federal funding sometimes serves as a means by which *private* research can be subjected to critical standards, since institutions that receive federal funds are often inclined (and given strong administrative incentives) to abide by the prescribed ethical standards throughout all of their activities, not only those directly receiving public dollars. Some supporters of funding therefore argue that extending public money to research is the most effective means of making certain that nearly all researchers, public and private, adhere to basic standards of ethics and safety. Public funding also requires researchers to make their work available to the public and for critical review by their peers, and it may encourage some degree of responsibility not necessarily encouraged by commercial endeavors.[viii]

In addition to conditions attached to government funding of research, law sometimes erects specific limits on certain practices that might be medically beneficial. For example, we put limits on some practices that might offer life-saving benefits, such as the buying and selling of organs for transplantation, currently prohibited under the National Organ Transplant Act. Also, as in the present case, many Americans and their congressional representatives have moral reasons for opposing certain lines of research or clinical practice, for example those that involve the exploitation and destruction of human fetuses and embryos.

The two sides of the embryo research debate tend to differ sharply on the fundamental moral significance of the activity in question. One side believes that what is involved is morally abhorrent in the extreme, while the other believes embryo research is noble or even morally obligatory and worthy of praise and support. It would be very difficult for the government to find a middle ground between these two positions, since the two sides differ not only on what should or should not be done, but also on the moral premises from which the activity should be approached.

To this point, the federal government has pursued a policy whereby it does not explicitly prohibit embryo research but also does not officially condone it, encourage it, or support it with public funds (though state governments have often taken more active roles in both directions, as detailed in Appendix E). This approach, again, combines prudential demands with moral concerns. It has allowed the political system to avoid banning embryo research against the wishes of those who believe it serves an important purpose, while not compelling those citizens who oppose it to fund it with their tax money. This approach is also based, at least in part, on the conviction that debates over the federal budget are not the place to take up the anguished question of the moral status of human embryos.

But the position is not only a compromise between those who would have the government bless and those who would have the government curse this activity. It is also a statement of a certain principle: namely, that public sanction makes a serious difference and ought not to be conferred lightly. While embryo destruction may be something that some Americans support and engage in, it is not something that America as a nation has officially supported or engaged in.[ix]

Of course, if the funding issue were merely a proxy for the larger dispute over the moral status of human embryos, then the present arrangement might appeal only to those who would protect human embryos, and it would succeed only as long as they were able to enact it. The argument might end there, with a vote-count on the question of the moral status or standing of human embryos. But some proponents of the present law suggest that the particulars and contours of the embryo research debate offer an additional rationale for that arrangement. Here again, it is important to remember that the issue in question is public funding, not permissibility. Opponents of embryo research have in most cases acquiesced (likely owing to various prudential and moral factors) in narrowing the debate at the federal level to the question of funding. They do not argue for a wholesale prohibition of embryo research by national legislation, even though many of them see such work as an abomination and even a form of homicide. In return, proponents of the Dickey Amendment argue that it would be appropriate for supporters of research to agree to do without federal funding in this particular field.

On the other hand, it might reasonably be argued that part of living under majority rule is living with the consequences of sometimes being in the minority. Were the Congress to overturn the current policy of withholding public funds from the destruction of embryos, opponents of funding for embryo research would not be alone in being compelled to pay for activities they abhor. We all see our government do things, in our name, with which we disagree. Some of these might even involve life and death questions of principle, for instance in waging wars that some citizens deeply oppose. The existence of strong moral opposition to some policy is not in itself a decisive argument against proceeding with that policy.

These concerns give the question of funding its own crucial ethical significance, even apart from the more fundamental question of the legitimacy and propriety of the act being funded. This matter of funding broadly understood, together with the moral and prudential aims apparently motivating the administration's policy, as well as the legal context created by the Dickey Amendment, are the essential prerequisites for thinking about the underlying logic of the current policy. The combination of these elements gives form not only to the specific rules set forth in the administration's funding policy, but also to the implementation of that policy, to which subject we now turn.

V. IMPLEMENTATION OF THE PRESENT POLICY

The complex and critical task of implementing the funding policy falls largely to the National Institutes of Health, which administers most federal funding of biomedical research. As noted, the administration's policy attempts to advance stem cell research within the bounds already laid out regarding further destruction of human embryos. Thus, while the funding criteria of the policy set the bounds, the NIH, in its ongoing work, is expected to advance the goal of maximally effective funding and support within those bounds.

To this end, the NIH has worked to "jump-start" this field of research through a series of coordinated activities.[20] To plan and oversee these activities, the NIH has established a Stem Cell Task Force charged with determining the best uses for public funds in the field and with putting in place the resources required to make effective use of those funds.

The most basic material resources in question are the human embryonic stem cell lines themselves. In August 2001, President Bush announced that "more than sixty genetically diverse stem cell lines" (or stem cell preparations) already existed, and so would be eligible for funding under his policy.[21] The NIH now believes the actual number to be somewhat higher, so that seventy-eight lines (or preparations) are known to be eligible for funding.[x] The lines are held by universities, companies, and other entities throughout the world. According to the National Institutes of Health's latest report (September 2003), the following organizations have developed stem cell derivations eligible for federal funding (that is, derived prior to August 9, 2001, under the approved conditions):

Name	Number of Derivations
Bresa Inc., Athens, Georgia	4
CyThera, Inc., San Diego, California	9
ES Cell International, Melbourne, Australia	6
Geron Corporation, Menlo Park, California	7
Göteborg University, Göteborg, Sweden	19
Karolinska Institute, Stockholm, Sweden	6
Maria Biotech. Co. Ltd., Maria Infertility Hospital Medical Institute, Seoul, Korea	3
MizMedi Hospital—Seoul National University, Seoul, Korea	1
National Centre for Biological Sciences/Tata Institute of Fundamental Research, Bangalore, India	3
Pochon CHA University, Seoul, Korea	2
Reliance Life Sciences, Mumbai, India	7
Technion University, Haifa, Israel	4
University of California, San Francisco	2
Wisconsin Alumni Research Foundation, Madison, Wisconsin	5

Although all of these lines (or preparations) are deemed *eligible* for funding according to the criteria of the administration's policy, not all are presently *available* for use by researchers (nor is it clear that *all* of them will ever be available for widespread use). Indeed, a point critical to understanding the current situation is that as of the autumn of 2003 only *twelve* lines are available for use,[22] while most of the other lines are not yet adequately characterized or developed (some exist only as frozen stocks) and so have at least not yet become available.[xi] The process of establishing a human embryonic stem cell line, turning the originally extracted cells into stable cultured populations suitable for distribution to researchers, involves an often lengthy process of growth, characterization, quality control and assurance, development, and distribution. In addition, the process of making lines available to federally funded researchers involves negotiating a contractual agreement (a "materials transfer agreement") with the companies or institutions owning the cell lines, establishing guidelines for payment, intellectual property rights over resulting techniques or treatments, and other essential legal assurances between the provider and the recipient.

The entire process—scientific and legal—has tended to take at least a year for each cell line. Thus, determining which of the 78 eligible lines are in sufficiently good condition, characterizing and developing those lines, and establishing the arrangements necessary to make them available has been a demanding task. By September of 2003, slightly over two years after the enactment of the funding policy, twelve of the eligible lines had become available to federally funded researchers.[xi] The NIH has made available "infrastructure award" funds (totaling just over $6 million to date) to a number of the institutions that possess eligible cell-lines, to enable them to more quickly and effectively develop more lines to distribution quality. As a result, while the number of available lines (only one in the summer of 2002 but risen to twelve in the autumn of 2003)[xi] is expected to continue to grow with time, it is unclear how many of the 78 lines will finally prove accessible and useful. According to the NIH, as of the autumn of 2003, the owners of the available lines have distributed over 300 shipments of lines to researchers. No information is presently available on the number of individual researchers or institutions that have received lines.[23]

Successful implementation of the current funding policy depends not only on the availability of eligible lines, but also on adequate allocation of financial resources to develop and make use of those lines and to advance the field in general. The funding policy, though it limits the targets of funding to the eligible lines, does not directly delimit or restrict the *amount of money* and other resources that the NIH may invest in human embryonic stem cell research. The amount invested, a decision left to NIH and the Congressional appropriations process, is largely a function of the number of qualified applicants for funding and of the NIH's own priorities and funding decisions. Of course, if more lines were eligible for funding, it is quite possible that more funding would be allocated, but the *amount* that *can* be allocated to work on existing lines is not limited by the funding criteria. In fiscal year 2002, the NIH devoted approximately $10.7 million to human embryonic stem cell research. Based upon a September 2003 estimate, it will have spent approximately $17 million in fiscal year 2003. It is expected that further increases will follow as the field and the number of grant applications grow.

As of November 2003, NIH funds have been allocated to support the following new and continuing awards for human embryonic stem cell research: nine infrastructure awards to assist stem cell providers to expand, test, and perform quality assurance, and improve distribution of cell lines that comply with the administration's funding criteria (aimed at making more of the eligible lines available); 28 investigator-initiated awards for specific projects; 88 administrative supplements (awarded to scientists already receiving funds for work on other sorts of stem cells, either non-embryonic or non-human, to enable them to begin to work with eligible human embryonic stem cell lines); two pilot and feasibility awards; three awards to support exploratory human embryonic stem cell centers; one institutional development award; four post-doctoral training fellowships; one career enhancement award; and six awards to fund stem cell training (including short-term courses) to provide hands-on training to enable researchers to learn the skills and techniques of culturing human embryonic stem cells.

The latter task, of training new researchers, the NIH regards as one of its principal challenges in advancing the field, and, along with available lines and available financial resources, as a key measure of how the field is progressing. As NIH Director Elias A. Zerhouni put it in his presentation before this Council, "I don't think the limiting factor is the cell lines. I really don't. I really think the limiting factor is human capital and trained human capital that can quickly evaluate a wide range of research avenues in stem cells."[24]

The NIH has therefore devoted funding to the training of investigators and the cultivation of career development pathways, including short-term courses in stem cell culture techniques and (long-term) career enhancement awards in the field. Some critics have contended, however, that the two issues (funding restrictions and the scarcity of personnel) are likely connected, and that limits on the cell lines eligible for funding and the surrounding political controversy cause some potential researchers to stay away from the field, contributing to a shortage of investigators.[25]

These federal resources, then, have been directed toward the advancement of human embryonic stem cell research within the bounds of the determination to refrain from supporting or funding new destruction of human embryos. Scientists may receive federal funding—at any level determined appropriate by the NIH—for any sort of meritorious research, using as many of the approved lines as they are eventually able to use. They can, of course, also receive federal funding for using or deriving new animal embryonic stem cell lines, to assess the potential of these cells for treatment of animal models of human disease (though of course animal models provide only limited information because they are not in many cases exactly extrapolatable to the specific situations that hold in human disease and development, and so cannot replace human cell sources).

Researchers can, in addition, use federal funds for work involving human embryonic germ cells, obtained from aborted fetuses. They can carry out research projects using embryonic germ cell lines already derived, following review and approval of specific institutional assurances, informed consent documents, scientific protocol abstracts, and Institutional Review Board approvals by the NIH's Human Pluripotent Stem Cell Review Group (HPSCRG). They can also receive federal funds for the derivation and study of *new* embryonic germ cell lines following the same HPSCRG review and approval process. In addition, of course, they can develop animal embryonic germ cell lines to assess the potential of these cells through animal models.

Also, researchers can receive federal funds for work conducted on human adult (non-embryonic) stem cells. There are no restrictions regarding what American scientists can do with regard to adult stem cells using taxpayer funds, other than those requiring them to honor the usual human subject protections and clinical research requirements (if they are to be transplanted into human patients). The NIH has devoted substantial resources to the study of human adult stem cells, allocating over $170 million to the field in fiscal year 2002, and approximately $181.5 million in fiscal year 2003 (approximately ten times the amount devoted to human embryonic stem cell work).

Finally, researchers remain free to pursue work (including the derivation of new lines of embryonic stem cells) in the private sector, without government funding. Indeed, as discussed above, embryonic stem cells were first isolated and developed in the private sector, or in university laboratories using private sector funds, and no work in the field was publicly funded at all until 2001. Under present law, work supported by private funds can proceed without restriction. Under rules promulgated in the spring of 2002, such work does not need to be conducted in a separate laboratory, but a clear separation of the funds used to support this work from any federally funded work of the laboratory is required. Of course, because of the highly interlocking and complex nature of the various aspects of operating a laboratory, such separation can still prove extremely difficult to manage. It is not clear precisely how much privately funded work using human embryonic stem cells has been undertaken in the past few years, but some general figures are available. The most recent and thorough survey available, based on figures from 2002,[26] suggests that approximately 10 companies in the United States were actively engaged in embryonic stem cell work, employing several hundred researchers and, cumulatively over the past several years, spending over $70 million in the field, which is well over twice what the NIH has so far spent.[27] Those involved in privately-funded research in the field, however, generally do not see private funding as a substitute for federal funds, but would much prefer that the field had the opportunity to benefit from both. They also argue that restrictions on federal funds, and the controversy surrounding the subject, act to dissuade potential investors from entering the field, and thereby have a "chilling effect" on private as well as publicly funded research.[28]

Moreover, just as federal policy can affect privately conducted research, so too a number of states have enacted policies affecting stem cell research, ranging from all-out prohibitions of such research to official statements of support and positive encouragement.[xii] The status of such research, and the conditions to which it is subject, can vary dramatically from state to state, independent of federal funding policy.

VI. CONCLUSION

The administration's policy on the funding of embryonic stem cell research rests on several moral and ethical-legal principles, set upon the reality of existing law:

1. *The law:* The Dickey Amendment, which the President is required to enforce.
2. *The principle underlying the law:* The conviction, voiced by the administration, a majority of the Congress, and some portion of the public, that federal taxpayer dollars should not be used to encourage the exploitation or destruction of nascent human life, even if scientific and medical benefits might come from such acts.
3. *The principle underlying the desire to offer funding:* That efforts to heal the sick and the injured are of great national importance and should be vigorously supported, provided that they respect important moral boundaries.

4. *The significance of federal funding:* That federal funding constitutes a meaningful positive statement of national approval and encouragement, which should be awarded only with care, particularly in cases where the activity in question arouses significant public moral opposition.

The significance of the policy is best understood in light of these key elements. Its soundness is most reasonably measured against them and against the policy's implementation by the National Institutes of Health. Though the prudential and principled considerations raised in this chapter governed the formulation of the policy, or at least defined its articulation by its advocates and authors, these are not the only terms by which federal funding policy might be conceived or measured. In the next chapter we present an overview of the ethical and policy debates that have raged for the past two years around both the wisdom of the present policy and the fundamental issues at stake in human embryonic stem cell research.

NOTES

i. Some Members of the Council oppose the current policy, some Members support it. Yet the descriptive account that we offer here aspires to be seen as accurate and fair, regardless of where one personally stands on the issue. Nearly all Members of this Council recognize, as we said in our report *Human Cloning and Human Dignity,* that "all parties to this debate have something vital to defend, something vital not only to themselves but also to their opponents in the debate, and indeed to all human beings. No human being and no human society can afford to be callous to the needs of suffering humanity, cavalier regarding the treatment of nascent human life, or indifferent to the social effects of adopting in these matters one course of action rather than another." (*Human Cloning and Human Dignity,* p. 121.) Thus, whatever we think of the current funding policy, we recognize that this is a genuine ethical dilemma and that reasonable people of good will may come to different conclusions about where the best ethical or policy position lies. We therefore also believe that not only results but also reasons matter, and that it behooves us to understand the principled or prudential reasons for the current policy (as well as for any alternative policy that might be offered to replace it).

ii. In its discussion of "fetal" research, the commission defined the fetus as the product of conception from the time of implantation onward, which therefore included what we generally think of (and define in this report) as embryos in utero. Its separate consideration of embryo research was therefore directed at in vitro embryos.

iii. These legal citations refer to the federal regulations and federal statutes relating to research on living human fetuses.

iv. These regulations, as published in the Federal Register, are provided in Appendix D.

v. The official NIH statement of this policy is provided in Appendix C.

vi. Using similar language, but speaking even more unambiguously, President Bush reiterated his ethical view of the destruction of human embryos for medical research in a speech on human cloning legislation, saying, "I believe all human cloning is wrong, and both forms of cloning ought to be banned, for the following reasons. First, anything other than a total ban on human cloning would be unethical. Research cloning would contradict the most fundamental principle of medical ethics, that no human life should be exploited or extinguished for the benefit of another. Yet a law permitting research cloning, while forbidding the birth of a cloned child, would require the destruction of nascent human life." ("Remarks by the President on Human Cloning Legislation," as made available by the White House Press Office, April 10, 2002.)

vii. Readers should note that in reporting on this approach, as applicable to President Bush's stem cell decision, the Council is not itself declaring its own views on whether the past act of embryo destruction was "immoral." (Some of us think it was, some of us think it wasn't.) We are rather describing what we understand to be the moral logic of the decision as put forward.

viii. Indeed, some even argue that the terms and conditions set for federal funding of research could be defined in such a way as not only to subject private research to general standards but also to help influence the eventual distribution of the products of that research to all those in need, or to serve other goods deemed publicly worthy.

ix. The repeated reenactment of the Dickey Amendment by the Congress may be taken as evidence of some support for this assertion.

x. These numbers took almost everyone by surprise. Prior to the President's announcement, the best estimates of the number of human embryonic stem cell lines then existing worldwide ranged between 10 and 20. But eligibility is not the same thing as availability, as we will discuss.

xi. By the time of final publication of this document, in January 2004, the number of available lines had risen to 15. This number is likely to rise further, and readers are advised to keep abreast of the current number and availability of embryonic stem cell lines eligible for funding at the NIH Stem Cell Registry website, stemcells.nih.gov.

xii. State policies regarding embryo research are detailed in Appendix E.

ENDNOTES

1. National Research Act, Pub. L. No. 93-348, § 213, 88 Stat. 342 (passed by the 93rd Congress as H.R. 7724, July 12, 1974).

2. National Commission for the Protection of Human Subjects of Biomedical and Behavioral Research, *Research on the Fetus: Report and Recommendations* (Washington, D.C., 1975). Reprinted at 40 Fed. Reg. 33,526 (1975).

3. "HEW Support of Human In Vitro Fertilization and Embryo Transfer: Report of the Ethics Advisory Board," 44 Fed. Reg. 35,033 (June 18, 1979) at 35,055–35,058.

4. Ibid.

5. National Institutes of Health Revitalization Act of 1993, Pub. L. No. 103-43, § 121(c), 107 Stat. 122 (1993) repealing 45 C.F.R. § 46.204(d).

6. National Institutes of Health, *Report of the Human Embryo Research Panel* (Bethesda, MD: NIH, 1994).

7. "Statement by the President," as made available by the White House Press Office, December 2, 1994.

8. The text of the Dickey Amendment can be found in each year's Labor/HHS Appropriations Bill. The original version, introduced by Representative Jay Dickey, is in § 128 of Balanced Budget Downpayment Act, I, Pub. L. No. 104-99, 110 Stat. 26 (1996). For subsequent fiscal years, the rider is found in Title V, General Provisions, of the Labor, HHS and Education Appropriations Acts in the following public laws: FY 1997, Pub. L. No. 104-208; FY 1998, Pub. L. No. 105-78; FY 1999, Pub. L. No. 105-277; FY 2000, Pub. L. No. 106-113; FY 2001, Pub. L. No. 106-554; and FY 2002, Pub. L. No. 107-116. The most current version (identical in substance to the rest) is in Consolidated Appropriations Resolution, 2003, Pub. L. No. 108-7, 117 Stat. 11 (2003)·

9. "Rendering legal opinion regarding federal funding for research involving human pluripotent stem cells," Memo from Harriet S. Rabb, General Counsel of the Department of Health and Human Services to Harold Varmus, Director of the National Institutes of Health, January 15, 1999. (Available through the National Archives.)

10. This case was made, for instance, in a letter authored by Rep. Jay Dickey and signed by seventy other Members of Congress to DHHS Secretary Donna Shalala, February 11, 1999.

11. DHHS Secretary Shalala argued this point in a letter responding to the Congressional letter of opposition (see note 10, above), February 23, 1999.

12. President Bush has made a number of statements articulating the position that nascent human life (including at the early embryonic stage) is deserving of protection and ought not be violated. See especially: "Stem Cell Science and the Preservation of Life," *The New York Times,* August 12, 2001, p. D13; "Remarks by the President on Human Cloning Legislation," as made available by the White House Press Office, April 10, 2002; "Remarks by the President at the Dedication of the Pope John Paul II Cultural Center," as made available by the White House Press Office, March 22, 2001; and "President Speaks at 30th Annual March for Life on the Mall," as made available by the White House Press Office, January 22, 2003.

13. "Remarks by the President on Stem Cell Research," as made available by the White House Press Office, August 9, 2001.

14. "Remarks by the President upon Departure for New Jersey," as made available by the White House Press Office, August 23, 2000.

15. National Institutes of Health, *Report of the Human Embryo Research Panel,* (Bethesda, MD: NIH, 1994).

16. See, for instance, the "Moral Case for Cloning-for-Biomedical-Research," presented by some Members of the Council in the Council's July 2002 report *Human Cloning and Human Dignity: An Ethical Inquiry*, chapter 6.

17. Bush, G.W., "Stem Cell Science and the Preservation of Life," *The New York Times,* August 12, 2001, p. D13.

18. See, for instance, Council discussion at its September 3, 2003, meeting. A transcript of that session is available on the Council's website at www.bioethics.gov.

19. This question has been addressed by the Supreme Court on a number of occasions, in which the Court found that even activities protected as rights under the Constitution are not thereby inherently worthy of financial support from the federal government. See, for instance, *Maher v. Roe* 432 U.S. 464 (1977); *Harris v. McRae* 448 U.S. 297 (1980); and *Rust v. Sullivan* 500 U.S. 173 (1991). Also see Berkowitz, P., "The Meaning of Federal Funding," a paper commissioned by the Council and included in Appendix F of this report.

20. The information provided in this section relies primarily on a presentation delivered before the Council by NIH Director Elias Zerhouni on September 4, 2003, and on data otherwise made available by the National Institutes of Health. The full transcript of Director Zerhouni's presentation may be found on the Council's website at www.bioethics.gov.

21. "Remarks by the President on Stem Cell Research," as made available by the White House Press Office, August 9, 2001.

22. As of the autumn of 2003, the following providers have eligible lines available for distribution: BresaGen (2 available lines), ES Cell International, Australia (5 available lines), MizMedi Hospital, South Korea (1 available line), Technion University, Israel (2 available lines), University of California at San Francisco (1 available line), Wisconsin Alumni Research Foundation (1 available line). A complete list of available and eligible lines, updated as more lines become available, can be found at the NIH Stem Cell Registry website at stemcells.nih.gov.

23. This information has been made available to the Council by the National Institutes of Health.

24. Quoted from a presentation before the Council by NIH Director Elias Zerhouni, September 4, 2003. The full transcript of Director Zerhouni's presentation maybe found on the Council's website at www.bioethics.gov.

25. See, for instance, the presentation of Thomas Okarma, President and CEO of Geron Corporation, before the Council on September 4, 2003. The full transcript of Okarma's presentation may be found on the Council's website at www.bioethics.gov.

26. Lysaght, M.J., and Hazlehurst, A.L., "Private Sector Development of Stem Cell Technology and Therapeutic Cloning," *Tissue Engineering* 9(3): 555–561 (2003).

27. The dollar amount spent specifically on embryonic stem cell research in the private sector is not apparent from Lysaght and Hazlehurst's survey. The $70 million figure is drawn from a presentation before the Council by Thomas Okarma, President and CEO of Geron Corporation, the oldest and largest of the private companies involved in embryonic stem cell research. Okarma told the Council, speaking only of Geron, "We have spent over $70 million on this technology, most of it since 1999 af-

ter the cells were derived. That's a number against which the NIH disbursements pale by both absolute and relative terms." The full transcript of Okarma's presentation may be found on the Council's website at www.bioethics.gov.

28. Ibid., and see also (for instance) Mitchell, S., "U.S. stem cell policy deters investors," *The Washington Times,* November 2, 2002 (original source: UPI).

11

Nuclear Transplantation, Embryonic Stem Cells, and the Potential for Cell Therapy

Konrad Hochedlinger and Rudolf Jaenisch

Nuclear cloning, also referred to as nuclear transfer or nuclear transplantation, denotes the introduction of a nucleus from an adult donor cell into an enucleated oocyte to generate a cloned embryo. When transferred to the uterus of a female recipient, this embryo has the potential to grow into an infant that is a clone of the adult donor cell, a process termed "reproductive cloning." However, when explanted in culture, this embryo can give rise to embryonic stem cells that have the potential to become any or almost any type of cell present in the adult body. Because embryonic stem cells derived by nuclear transfer are genetically identical to the donor and thus potentially useful for therapeutic applications, this process is called "nuclear transplantation therapy" or "therapeutic cloning." Therapeutic cloning might substantially improve the treatment of neurodegenerative diseases, blood disorders, or diabetes, since therapy for these diseases is currently limited by the availability or immunocompatibility of tissue transplants. Indeed, experiments in animals have shown that nuclear cloning combined with gene and cell therapy represents a valid strategy for treating genetic disorders.

Reproductive cloning is an inefficient and error-prone process that results in the failure of most clones during development. For a donor nucleus to support development, it must properly activate genes important for early embryonic development and suppress differentiation-associated genes that were transcribed in the original donor cell. Inadequate "reprogramming" of the donor nucleus is thought to be the principal reason for the developmental loss of most clones. In contrast, reprogramming errors do not appear to interfere with therapeutic cloning, because the process appears to select for functional cells.

Recent advances in the field of nuclear cloning allow four major conclusions to be drawn. First, most clones die early in gestation, and only a few survive to birth or beyond. Second, cloned animals have common abnormalities regardless of the type of donor cell or the species used, and third, these abnormalities correlate with aberrant gene expression, which most likely results from faulty genomic reprogramming. Fourth, the efficiency of cloning depends on the state of differentiation of the donor cell. In this article, we will summarize recent results from our laboratory and those of others and review potential therapeutic applications of the nuclear-cloning technology.

THE STATE OF THE ART OF NUCLEAR CLONING

Common Abnormalities in Cloned Animals

Most cloned embryos die soon after implantation.[1–3] Those that live to birth often have common abnormalities irrespective of the type of donor cell used (Table 11.1). For instance newborn clones are frequently unusually large and have an enlarged placenta (the large offspring syndrome).[2, 7, 10, 14–17] Moreover, neonate clones often have respiratory distress and defects of the kidneys, liver, heart, and brain.[18] Even long-term survivors can have abnormalities later in life. Aging cloned mice were recently reported to become obese,[19] die prematurely, and have tumors.[20]

Some of these phenotypes do, however, appear to be specific to the type of donor cell used. For example, clones derived from cumulus cells (somatic cells that surround the egg) become obese,[19] whereas clones derived from Sertoli cells (somatic cells that nourish the egg) die prematurely.[20] However, these abnormalities are not inherited by the offspring of the clones, suggesting that epigenetic rather than genetic aberrations are the cause; epigenetic changes, in contrast to genetic changes, are reversible modifications of DNA or chromatin that are usually erased in the germ line. These results indicate that most problems associated with cloning appear to be due to faulty epigenetic reprogramming of the transplanted nucleus.

Faulty Epigenetic Reprogramming in Clones

Faulty epigenetic reprogramming is the failure to return the gene-expression program of a somatic donor nucleus to an embryonic pattern of expression.[2] At the molecular level, epigenetic modifications that are specific to the differentiated cell, such as DNA methylation, histone modifications, and the overall chromatin structure, need to be reprogrammed to a state compatible with embryonic development. Consistent with this notion is the finding that embryos cloned from somatic cells frequently fail to reactivate key embryonic genes at the blastocyst stage.[21, 22] Moreover, cloned embryos can have aberrant patterns of DNA methylation[23–25] and precocious expression of genes specific to the donor cell.[26] In contrast, embryos cloned from embryonic stem cells faithfully express early

Table 11.1. Characteristics of Cloned Mice*

| | Developmental Characteristic | | | | |
| Type of Donor Cell | Zygote into Blastocyst | Blastocyst into Mouse | Blastocyst into Embryonic Stem Cells | Phenotype of Clones | References |
		(percentage of clones)			
B or T cells	4	0†	7	Not analyzed	Hochedlinger and Jaenisch[5] and unpublished data
Fibroblasts	40–60	1	3–6	Abnormal‡	Wakayama et al.[6] and Wakayama and Yanagimachi[7]
Cumulus cells	40–70	1–3	10–14	Abnormal‡	Wakayama et al.[6, 8] and Munsie et al.[9]
Blastomeres or embryonic stem cells	10–15	10–30	Not analyzed	Abnormal‡	Eggan et al.,[10] Rideout et al.,[11] and Cheong et al.[12]
Zygote pronuclei	90–100	20–40	30–60	Normal	McGrath and Solter[13] and unpublished data

*Shown is a comparison of preimplantation ("zygote into blastocyst") and postimplantation ("blastocyst into mouse") development of clones derived from different donor cells. The use of embryonic cells (blastomere cells and embryonic stem cells) leads to a higher rate of postimplantation development of cloned embryos than does the use of adult donor cells (fibroblasts and cumulus cells). Therapeutic cloning ("blastocyst into embryonic stem cells") is more efficient than reproductive cloning ("blastocyst into mouse"). Regardless of the type of donor cell used, cloned animals have common abnormalities.

†No live cloned mice have been derived from lymphocytes by direct embryo transfer.[4] However, monoclonal mice have been generated by a modified, two-step cloning procedure.[5]

‡Abnormalities include an unusually large fetus and placenta, respiratory distress, and defects of the kidneys, liver, heart, and brain.

embryonic genes,[21] possibly because these genes are already active in the donor genome. This might explain why cloning from embryonic stem cells is roughly 10 to 20 times as efficient as cloning from somatic cells (Table 11.1).

During normal development, reprogramming occurs before and after the formation of the zygote[2, 27] (Table 11.2). Faithful reprogramming ensures the proper activation of genes during embryonic development. Prezygotic reprogramming includes the acquisition of genomic imprints—the expression of genes from either the paternal or maternal set of chromosomes—as well as the modification of most somatic genes during gametogenesis. Inactivation of the X chromosome and adjustment of the length of telomeres are examples of postzygotic reprogramming.

Prezygotic Reprogramming

Because cloning uses an unfertilized, mature oocyte, reprogramming has to occur within the brief interval between the transfer of the donor nucleus into the oocyte and the start of zygotic transcription. Thus, prezygotic modifications (i.e., any modifications that have occurred before the mature oocyte stage) are expected to be less efficiently reprogrammed than postzygotic mod-

Table 11.2.   Outcome of Epigenetic Reprogramming in Cloned Animals*

Timing of Normal Reprogramming	Epigenetic Modification	Outcome of Reprogramming in Clones	References
Before the zygote is formed (during gametogenesis)	Modification of non-imprinted genes	Faulty in 4% of tested genes	Humpherys et al.[28]
	Acquisition of gene imprinting	Faulty in 20–50% of tested genes	Humpherys et al.[17]
After the zygote is formed	X-inactivation	Same as in normal embryos (random in the mouse)	Eggan et al.[29]
	Adjustment of telomere length	Same as in normal embryos (ends are normal or longer)	Betts et al.,[30] Lanza et al.,[31] Tian et al.,[32] and Wakayama et al.[33]

*During normal development, epigenetic reprogramming occurs before and after the formation of the zygote to ensure activation of the proper genes in the developing embryo. During clonal development, reprogramming is limited to the short interval between the transfer of the nucleus into the mature oocyte and the activation of embryonic transcription. Consequently, prezygotic modifications of imprinted and non-imprinted genes are less efficiently reprogrammed in clones than are postzygotic modifications, such as inactivation of the X chromosome and adjustment of the length of telomeres. Data were adapted from Jaenisch et al.[27] with the permission of the publisher; gene expression of imprinted and nonimprinted genes was analyzed with use of 11K Affymetrix gene chips.

ifications. Consistent with this hypothesis is the fact that aberrant imprints in donor nuclei are usually not corrected in the clones. Genomic imprinting is an epigenetic modification of DNA resulting in the mono-allelic and parent-of-origin-specific expression of certain genes. The dysregulation of imprinted genes is particularly pronounced in cloned mice derived from embryonic stem cells, because cultured embryonic stem cells are epigenetically very unstable and frequently gain or lose genomic imprints.[16] Cloned mice derived from un-cultured cumulus cells with normal imprints also have aberrant expression of imprinted genes,[28] suggesting that the dysregulation of imprinted genes is influenced by both the epigenetic state of the donor cell and the nuclear-transfer procedure. Since imprinted genes are important for fetal growth and placental function, aberrant expression of these genes might account for the severely abnormal fetal and placental phenotypes in many clones.

To assess the degree of dysregulation of nonimprinted genes, gene-expression analyses have been performed on newborn clones. These analyses revealed that hundreds of genes are aberrantly expressed in the placentas and livers of cloned mice derived from either cumulus or embryonic stem cells[28] (Table 11.2). Interestingly, a subgroup of these genes was found to be misexpressed exclusively in clones derived from cumulus cells, and another subgroup was aberrantly expressed only in clones derived from embryonic stem cells—a result consistent with the finding that clones derived from different types of donor cells can have different abnormalities.[3, 19, 20] Therefore, prezygotic reprogramming, which affects the expression of imprinted and most nonimprinted genes, appears to be faulty in clones.

Postzygotic Programming

Telomeres are structures that protect the ends of chromosomes. Telomeres progressively shorten with each cell division, and this shortening has been correlated with cellular and organismal aging. In most cloned animals, the lengths of telomeres are normal or even longer than normal,[30–33] suggesting that cloned embryos faithfully restore telomere length to that of normal embryos (Table 11.2).

Inactivation of one X chromosome in female cells is a mechanism that ensures equal dosage of X-linked genes in the two sexes. It is accomplished by the random and stable silencing of one of the two X chromosomes early in embryogenesis. In embryos cloned from female somatic cells, the inactive X was found to be reactivated properly, resulting in random X inactivation in the mouse[29] (Table 11.2). Thus, most postzygotic modifications appear to be properly reprogrammed in clones and are therefore not expected to impede the development of clones.

In summary, all available evidence indicates that reproductive cloning, in contrast to normal development or in vitro fertilization, is limited by the fundamental biologic problem of epigenetic reprogramming of the donor nucleus. Specifically, prezygotic modifications that usually occur during gametogenesis are not corrected in the clones. This incomplete reprogramming may result in abnormal phenotypes, aberrant gene expression, and the death of most clones. Consequently, even the rare surviving clones are likely to have at least subtle abnormalities.

Differentiation and Cloning Efficiency

The efficiency of obtaining cloned animals from adult donor cells is low in most species. In general, 1 to 3 percent of cloned blastocysts develop completely[1, 2, 7, 8] (Table 11.1). This rate is slightly higher in cows, in which up to 10 percent of cloned embryos develop to term.[34] In contrast to the results of cloning involving somatic donor cells, the results of cloning involving embryonic cells such as blastomeres (cells of the cleavage embryo) or embryonic stem cells are more efficient (10 to 30 percent of cloned blastocysts develop successfully)[3,10–12] (Table 11.1), suggesting that the state of differentiation of the donor cell directly affects the efficiency of cloning. This observation is consistent with the idea that embryonic cells require less reprogramming of their genome, because the genes essential for early embryonic development are already active. In fact, the transfer of the nucleus of an embryonic cell, such as an embryonic stem cell, might have nearly the same rate of success in generating live-born mice as does transfer of the zygotic nucleus (or pro-nucleus)[13] (Table 11.1). However, mice cloned from the nuclei of embryonic stem cells, in contrast to mice derived from zygotic nuclei, are abnormal,[10] indicating that gametogenesis and fertilization endow zygotic nuclei with the ability to direct normal development. In summary, these data indicate that cells progressively lose nuclear potency as they develop.

Terminally Differentiated Cells Remain Totipotent

The loss of nuclear potency in differentiating cells raised the important question of whether the nuclei of terminally differentiated donor cells retain developmental totipotency — that is, the potential to give rise to an entire organism. Previous nuclear-transfer experiments in frogs[35] and mammals[36] failed to resolve this question, because of the lack of appropriate genetic markers that would unambiguously identify terminally differentiated cells within a heterogeneous population of donor cells. It is possible that the clones were not derived from differentiated donor cells but rather from adult stem cells that were present in the adult donor animal at low frequencies.[3, 37–39]

To address this question, we created monoclonal mice from terminally differentiated lymphocytes.[5] The characteristic genetic rearrangements at the immune-receptor loci of mature lymphoid cells served as genetic markers, which allowed us to draw the retrospective and unambiguous conclusion that the clones had been derived from a terminally differentiated donor nucleus. We found that nuclei from mature B and T cells were able to direct development after being transferred into an oocyte, but this process was much less efficient than cloning involving other adult donor cells, such as fibroblasts or cumulus cells[6–8] (Table 11.1). Previous attempts to generate monoclonal mice from lymphoid donor nuclei by the direct transfer of blastocysts into the uterus were unsuccessful.[4] To derive monoclonal mice from the nuclei of mature B and T cells, we used a two-step cloning procedure in which the derivation of embryonic stem cells from cloned blastocysts was followed by tetraploid embryo complementation.[10] In this approach, diploid embryonic stem cells are injected into tetraploid host blastocysts to generate mice. Because tetraploid cells can form a functional placenta but not an embryo, the resultant mice had to have been derived entirely from the injected embryonic stem cells.

Although the generation of monoclonal mice demonstrated unequivocally that terminally differentiated cells can remain genetically totipotent, these results did not exclude the possibility that many cloned animals are derived from less well differentiated adult cells, such as adult stem cells. The genome of adult stem cells might resemble that of embryonic stem cells, which is more amenable to or requires less reprogramming than the genome of a differentiated cell. It will be interesting to test whether purified adult stem cells can serve as efficient somatic donor cells. This question is also of importance with respect to the potential therapeutic application of nuclear transfer; because nuclear transfer is inherently inefficient, it will be essential to identify the most efficient donor cell in the adult in order to reduce the number of oocytes that are needed to establish a line of embryonic stem cells.

THERAPEUTIC POTENTIAL OF NUCLEAR TRANSPLANTATION

Reproductive Cloning versus Therapeutic Cloning

In addition to its value in the study of nuclear changes during differentiation, nuclear-transfer technology has substantial therapeutic potential. For the following discussion, it is important to distinguish between reproductive cloning and nuclear-transplantation therapy (also referred to as therapeutic cloning)[40, 41] (Fig. 11.1). The purpose of reproductive cloning is to generate

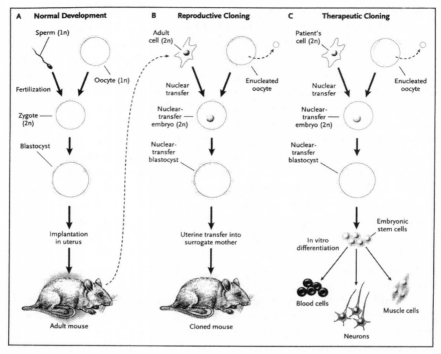

Figure 11.1. Comparison of Normal Development with Development during Reproductive Cloning and Therapeutic Cloning.
During normal development (Panel A), a haploid (1n) sperm cell fertilizes a haploid oocyte to form a diploid (2n) zygote that undergoes cleavage to become a blastocyst embryo. Blastocysts are implanted in the uterus and ultimately give rise to an animal. During reproductive cloning (Panel B), the diploid nucleus of an adult donor cell is introduced into an enucleated oocyte, which after artificial activation divides into a cloned blastocyst. On transfer into surrogate mothers, a few of the cloned blastocysts will give rise to a clone. In contrast, therapeutic cloning (Panel C) requires the explantation of cloned blastocysts in culture to yield a line of embryonic stem cells that can potentially differentiate in vitro into any type of cell for therapeutic purposes.

a cloned embryo, which is then implanted in the uterus of a female to give rise to a cloned individual. In contrast, the purpose of nuclear-transplantation therapy is to generate an autologous embryonic stem-cell line that is derived from a cloned embryo—referred to as nuclear-transfer embryonic stem cells—and that can be used for tissue replacement.

Rejection is a frequent complication of allogeneic organ transplantation, owing to immunologic incompatibility. To prevent this host-versus-graft disease, immunosuppressive drugs are routinely given to transplant recipients—a treatment that has serious side effects. Embryonic stem cells derived from nuclear transplantation are genetically identical to the patient's cells, thus eliminating the risk of immune rejection and the requirement for immuno-suppression. Moreover, embryonic stem cells provide a renewable source of replacement tissue, allowing therapy to be repeated whenever needed.

Differentiation into Functional Cells

Therapeutic cloning requires the in vitro differentiation of nuclear-transfer embryonic stem cells into a homogeneous population of functional cells that can be used for cell therapy. In some circumstances, these cells may first need to be manipulated to correct defects. Recently, protocols have been developed that allow homologous recombination and thus genetic manipulation of human embryonic stem cells.[42] Various studies have described the potential of human embryonic stem cells to differentiate into multiple lineages,[43, 44] such as neural progenitors,[45–47] hematopoietic precursors,[48] and insulin-secreting cells.[49]

Protocols for the differentiation of murine embryonic stem cells into functional cells of many if not all organs present in adult mice are well established (Table 11.3). For example, embryonic stem cells can generate functional motor neurons when they are exposed to signals that normally induce neurogenesis.[50] With a different strategy, drug-selection protocols have been used to cause embryonic stem cells to differentiate into homogeneous populations of cardiomyocytes,[54] neuroepithelial precursor cells,[55] and insulin-producing

Table 11.3. Examples of the Differentiation of Murine Embryonic Stem Cells into Functional Somatic Cells

Result of Differentiation	Protocol for Differentiation	Test of Functionality	References
Motor neurons	Stimulation by retinoic acid and sonic hedgehog	Integration into chicken spinal cord; muscle	Wichterle et al.[50]
Midbrain neurons	Constitutive expression of nuclear-receptor-related factor 1	Dopamine production and behavioral recovery in rat model of Parkinson's disease	Kim et al.[51]
Pancreatic islet-like cells	Selection for insulin-expressing cells	Insulin secretion and normalization of blood glucose levels in diabetic mice	Soria et al.[52]
Hematopoietic precursors	Transient expression of homeobox protein HoxB4	Myeloid and lymphoid engraftment in irradiated mice; transplantation into secondary recipients	Kyba et al.[53]

cells.[52] The expression of the homeobox protein HoxB4 in embryoid bodies generates hematopoietic stem cells that avert death in mice that have received lethal doses of radiation.[53] Similarly, the expression of nuclear receptor-related factor 1 (Nurr1), an orphan nuclear receptor that is expressed chiefly in the central nervous system, in embryonic stem cells induces the formation of dopaminergic neurons that can relieve behavioral symptoms in rats with Parkinson's disease.[51]

Combining Nuclear Cloning with Gene and Cell Therapy

The ultimate goal of therapeutic cloning is to generate functional cells from cloned embryonic stem cells that can be used for cell transplantation in patients. Several groups, including ours, have shown that nuclear-transfer embryonic stem cells can be derived from mouse cumulus or fibroblast cells and that these cells can be coaxed into becoming somatic cells such as myogenic cells, dopaminergic and serotonergic neurons, or hematopoietic cells.[5, 6, 9, 56] However, before these principles can be applied clinically, it is important to demonstrate the feasibility of therapeutic cloning in an animal model of disease.

In an attempt to establish such a mouse model, we have combined nuclear cloning with gene and cell therapy to treat a genetic disorder (Fig. 11.2).[56] We chose the well-characterized *Rag2* mutant mouse, which has severe combined immunodeficiency owing to a mutation in the recombination-activating gene 2 (*Rag2*), which catalyzes immune-receptor rearrangements in lymphocytes. This mouse is devoid of mature B and T cells, a condition resembling Omenn's syndrome in humans. First, we isolated somatic (fibroblast) cells from the tails of *Rag2* deficient mice and injected the nuclei of these cells into enucleated eggs. We then cultured the resultant embryos to the blastocyst stage and isolated the autologous embryonic stem cells. Subsequently, one of the mutant *Rag2* alleles was targeted by homologous recombination in embryonic stem cells in order to restore normal gene structure. To obtain somatic cells for treatment, these embryonic stem cells underwent differentiation into embryoid bodies (embryo-like structures that contain various types of somatic cells) and further into hematopoietic precursors by expressing HoxB4. The resulting hematopoietic precursors were transplanted into irradiated *Rag2*-deficient animals to treat the disease.

The initial attempts to engraft these cells were unsuccessful because of an increased level of natural killer cells in the mutant host. Hematopoietic cells derived from embryonic stem cells express low levels of major-histocompatibility-complex class I molecules and are thus a preferred target of destruction by natural killer cells. Elimination of the natural killer cells by

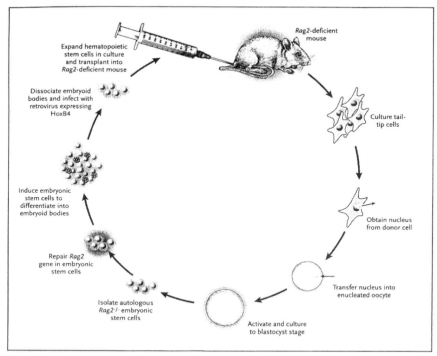

Figure 11.2. Mouse Model of Therapeutic Cloning.
Tail-tip cells were obtained from mice with a deficiency of recombination-activating gene 2 (*Rag2*) and cultured, and the nuclei were transferred into enucleated oocytes. The cloned embryos were cultured to the blastocyst stage to derive autologous embryonic stem cells. After one of the mutant *Rag2* alleles was repaired by homologous recombination, embryonic stem cells were induced to differentiate into embryoid bodies (embryo-like structures that contain various types of somatic cells) and infected with a retrovirus expressing the homeobox protein HoxB4. The resultant hematopoietic stem cells were clonally expanded and injected intravenously into irradiated *Rag2*-deficient animals to reconstitute their immune system. Adapted from Rideout et al.,[56] with the permission of the publisher.

antibody depletion or genetic ablation allowed the nuclear-transfer embryonic stem cells to differentiate into the myeloid lineages efficiently and, to a lesser degree, into the lymphoid lineages. Functional B and T cells whose immunoglobulin and T-cell-receptor alleles had been properly rearranged were detected in the mice, as were serum immunoglobulins. However, because HoxB4 appears to promote the differentiation of embryonic stem cells into myeloid cells, lymphoid reconstitution might be more successful if transcription factors specific to the lymphoid lineage were used.

This experiment demonstrated that nuclear transfer can be combined with gene therapy to treat a genetic disorder. Consequently, therapeutic cloning should be useful in other diseases in which the genetic cause is known, such as sickle cell anemia and thalassemia.

LIMITATIONS AND ALTERNATIVES

Faulty Reprogramming in Clones as a Potential Impediment to Therapeutic Application

An important question is whether the reprogramming errors leading to abnormal phenotypes in cloned animals would impede the therapeutic use of nuclear-transfer technology. For the derivation of embryonic stem cells from fertilized embryos, blastocysts are explanted in vitro and cultured until a small colony forms that can be dissociated. Only one or a few of the dissociated cells have the potential to grow into an embryonic stem-cell line,[57] suggesting that competent cells are selected for in culture. Similarly, the derivation of embryonic stem cells from cloned blastocysts may be the result of selection for a few successfully reprogrammed cells within a cloned embryo (Fig. 11.3). In contrast, the development of a cloned embryo after implantation most likely does not allow for the in vivo selection of a few functional cells, thus causing developmental failure of the clone or phenotypic abnormalities. In support of this notion, the derivation of embryonic stem cells from somatic donor cells is more efficient[5, 6, 9, 56] than is the generation of cloned mice[7, 8] (Table 11.1).

Abnormal fetal development is the most fundamental cause of clone failure. In contrast to the result of reproductive cloning, no fetus is formed in therapeutic cloning. Thus, aberrant expression of genes that are essential for normal fetal development, such as imprinted genes, is not expected to impede the functionality of embryonic stem cells that undergo differentiation in vitro. The abnormal expression of some imprinted genes, such as the gene for insulin-like growth factor 2, however, has been associated with disease in the adult[58, 59] and it will be important to determine whether dysregulation of these genes has adverse effects on the function of somatic cells derived from embryonic stem cells.

Nuclear-transfer embryonic stem cells have the same developmental potency as embryonic stem cells derived from fertilized eggs. When injected into blastocysts, nuclear-transfer embryonic stem cells contributed to tissues of all three germ layers, including the germ line.[6] A subgroup of these embryonic stem cells was even able to produce mice after tetraploid embryo complementation,[5, 56] a process that allows the generation of mice from em-

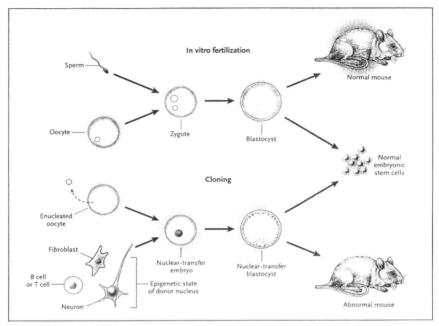

Figure 11.3. Derivation of Embryonic Stem Cells from a Blastocyst, Resulting in Selection for Functional Cells.
Only a few of the cells of a blastocyst derived from a fertilized zygote have the potential to produce an embryonic stem-cell line, and these cells seem to be selected for by culture conditions (top panel). Mice derived from fertilized zygotes are normal because the sperm and oocyte genomes have undergone proper reprogramming during gametogenesis (top panel). However, cloned blastocysts and the resultant mice seem to maintain a memory of the epigenetic state of the donor nucleus they were derived from (bottom panel). This is probably due to faulty reprogramming of the somatic donor nucleus after nucleus transfer and results in abnormal phenotypes and aberrant patterns of gene expression. These epigenetic abnormalities are represented by the pink halo surrounding the cloned embryo and mouse. In contrast, the derivation of embryonic stem cells from cloned blastocysts appears to select for fully reprogrammed, functional cells that have lost this epigenetic memory.

bryonic stem cells alone. It is notable that the abnormalities regularly associated with cloned animals were not observed in these mice.

In any therapeutic setting, cells derived from nuclear-transfer embryonic stem cells will be introduced into a patient with a disease, and host cells will interact with the transplanted cloned cells to generate a chimeric tissue. Chimeric animals generated by the injection of normal or nuclear-transfer embryonic stem cells into normal blastocysts form normal chimeras.[6, 9] This finding suggests that the presence of host helper-cells, which are derived from the fertilized egg, complement the defects that invariably result when a cloned animal in generated from a somatic or embryonic donor nucleus. In support of this notion, Byrne et al. showed that embryonic cells derived from nuclear transfer failed to develop on their own in frogs but integrated normally when they were combined with wild-type embryos to form chimeric tadpoles.[60] In summary, these considerations suggest that nuclear-transfer embryonic stem cells are equivalent to embryonic stem cells derived from a fertilized zygote.

Adult Stem Cells as an Alternative to Therapeutic Cloning

Are there alternatives to therapeutic cloning? Adult stem cells are another potential source of autologous cells for transplantation therapy. They have been isolated from adult tissues such as brain, bone marrow, skin, and muscle, and they might have a broader developmental potential than originally anticipated.[61] However, it remains unclear whether the observed plasticity, or "transdifferentiation potential," of adult stem cells is inherent to the cells or the consequence of culture conditions, contamination, or cell fusion.[61–65] Moreover, recent experiments have failed to reproduce the results of some earlier reports claiming that transdifferentiation occurred.[66, 67]

The therapeutic potential of adult stem cells appears to be much lower than that of embryonic stem cells. First, adult stem cells are difficult to isolate and hard to propagate in culture. In contrast, embryonic stem cells are derived rather easily (once an embryo has been obtained), and they grow indefinitely in culture. Second, embryonic stem cells can be manipulated genetically by homologous recombination to correct a genetic defect.[56] In contrast, currently, adult stem cells can be genetically manipulated only through the introduction of retroviral transgenes, which overexpress genes at variable levels and can lead to insertional mutagenesis and cancer.[68] Third, embryonic stem cells can be coaxed into becoming any type of cell through the use of specific culture conditions or genetic manipulation. The differentiation potential of adult stem cells, however, seems to be restricted.

One notable exception in this respect is the recent isolation of multipotent adult progenitor cells.[69] Multipotent adult progenitor cells were derived from the bone marrow of adult mice, rats, and humans after a three-month culture protocol. These cells have the potential to differentiate into cells of all three germ layers both in vitro and in vivo after being injected into blastocysts. However, it has not been demonstrated in animal models or humans that multipotent adult progenitor cells can be used to correct a disease phenotype.

The Requirement for Human Oocytes

To overcome the ethical and practical limitations of therapeutic cloning, it would be useful to reprogram somatic cells directly into embryonic stem cells without the use of oocytes. An understanding of the factors that have a role in establishing and maintaining pluripotency might make it possible to alter the fate of somatic cells directly. For instance, the embryonic transcription factor Oct-4 appears to act as a regulator of pluripotency during development.[57, 70] When mutated, embryos cannot form a pluripotent inner cell mass, and their development is arrested.[71] Thus, manipulation of Oct-4 and related genes[21] in somatic cells might help to reprogram their nuclei to an embryonic state. This could reduce or even circumvent the need for human oocytes.

Recently, Hübner et al. provided evidence of the differentiation of murine embryonic stem cells into oocyte-like cells in vitro.[72] It will be interesting to determine whether oocytes can be obtained from human embryonic stem cells and whether these cells are suitable for nuclear transfer.

CONCLUSION

Therapeutic cloning, in combination with the differentiation potential of embryonic stem cells, offers a valuable means of obtaining autologous cells for the treatment of a variety of diseases. The abnormalities associated with reproductive cloning are not expected to impede the use of this technique for therapy, since the process seems to select for functional cells. However, before these principles can be applied clinically, it will be essential to improve differentiation protocols for human embryonic stem cells and to evaluate the effect of oocyte-derived mitochondrial proteins in somatic cells obtained by nuclear transfer. In the future it might be possible to generate embryonic stem cells directly from somatic cells. It is important, therefore, to continue research aimed at improving our understanding of the molecular events that take place during nuclear reprogramming, in order to develop these potential new therapies.

NOTES

Supported by a Ph.D. fellowship from the Boehringer Ingelheim Funds (to Dr. Hochedlinger) and by a grant (R37-CA84198) to Dr. Jaenisch from the National Cancer Institute.

We are indebted to Caroline Beard, Robert Blelloch, Kevin Eggan, Joost Gribnau, and Teresa Holm for discussions and critical reading of the manuscript.

1. Solter D. Mammalian cloning: advances and limitations. Nat Rev Genet 2000;1:199–207.

2. Rideout WM III, Eggan K, Jaenisch R. Nuclear cloning and epigenetic reprogramming of the genome. Science 2001;293:1093–8.

3. Hochedlinger K, Jaenisch R: Nuclear transplantation: lessons from frogs and mice. Curr Opin Cell Biol 2002;14:741–8.

4. Wakayama T, Yanagimachi R. Mouse cloning with nucleus donor cells of different age and type. Mol Reprod Dev 2001;58:376–83.

5. Hochedlinger K, Jaenisch R. Monoclonal mice generated by nuclear transfer from mature B and T donor cells. Nature 2002;415:1035–8.

6. Wakayama T, Tabar V, Rodriguez I, Perry AC, Studer L, Mombaerts P. Differentiation of embryonic stem cell lines generated from adult somatic cells by nuclear transfer. Science 2001;292:740–3.

7. Wakayama T, Yanagimachi R. Cloning of male mice from adult tail-tip cells. Nat Genet 1999;22:127–8.

8. Wakayama T, Perry AC, Zuccotti M, Johnson KR, Yanagimachi R. Full-term development of mice from enucleated oocytes injected with cumulus cell nuclei. Nature 1998;394:369–74.

9. Munsie MJ, Michalska AE, O'Brien CM, Trounson AO, Pera MF, Mountford PS. Isolation of pluripotent embryonic stem cells from reprogrammed adult mouse somatic cell nuclei. Curr Biol 2000;10:989–92.

10. Eggan K, Akutsu H, Loring J, et al. Hybrid vigor, fetal overgrowth, and viability of mice derived by nuclear cloning and tetraploid embryo complementation. Proc Natl Acad Sci U S A 2001;98:6209–14.

11. Rideout WM III, Wakayama T, Wutz A, et al. Generation of mice from wild-type and targeted ES cells by nuclear cloning. Nat Genet 2000;24:109–10.

12. Cheong HT, Takahashi Y, Kanagawa H. Birth of mice after transplantation of early cell-cycle-stage embryonic nuclei into enucleated oocytes. Biol Reprod 1993;48:958–63.

13. McGrath J, Solter D. Completion of mouse embryogenesis requires both the maternal and paternal genomes. Cell 1984;37:179–83.

14. Tanaka S, Oda M, Toyoshima Y, et al. Placentomegaly in cloned mouse concepti caused by expansion of the spongiotrophoblast layer. Biol Reprod 2001;65:1813–21.

15. Hill JR, Burghardt RC, Jones K, et al. Evidence for placental abnormality as the major cause of mortality in first-trimester somatic cell cloned bovine fetuses. Biol Reprod 2000;63:1787–94.

16. Young LE, Sinclair KD, Wilmut I. Large offspring syndrome in cattle and sheep. Rev Reprod 1998;3:155–63.

17. Humpherys D, Eggan K, Akutsu H, et al. Epigenetic instability in ES cells and cloned mice. Science 2001;293:95–7.

18. Cibelli JB, Campbell KH, Seidel GE, West MD, Lanza RP. The health profile of cloned animals. Nat Biotechnol 2002;20:13–4.

19. Tamashiro KL, Wakayama T, Akutsu H, et al. Cloned mice have an obese phenotype not transmitted to their offspring. Nat Med 2002;8:262–7.

20. Ogonuki N, Inoue K, Yamamoto Y, et al. Early death of mice cloned from somatic cells. Nat Genet 2002;30:253–4.

21. Bortvin A, Eggan K, Skaletsky H, et al. Incomplete reactivation of Oct4-related genes in mouse embryos cloned from somatic nuclei. Development 2003;130:1673–80.

22. Boiani M, Eckardt S, Scholer HR, McLaughlin KJ. Oct4 distribution and level in mouse clones: consequences for pluripotency. Genes Dev 2002;16:1209–19.

23. Kang YK, Koo DB, Park JS, et al. Aberrant methylation of donor genome in cloned bovine embryos. Nat Genet 2001;28:173–7.

24. Kang YK, Park JS, Koo DB, et al. Limited demethylation leaves mosaic-type methylation states in cloned bovine preimplantation embryos. EMBO J 2002; 21:1092–100.

25. Dean W, Santos F, Stojkovic M, et al. Conservation of methylation reprogramming in mammalian development: aberrant reprogramming in cloned embryos. Proc Natl Acad Sci U S A 2001;98:13734–8.

26. Gao S, Chung YG, Williams JW, Riley J, Moley K, Latham KE. Somatic cell-like features of cloned mouse embryos prepared with cultured myoblast nuclei. Biol Reprod (in press).

27. Jaenisch R, Eggan K, Humpherys D, Rideout W, Hochedlinger K. Nuclear cloning, stem cells, and genomic reprogramming. Cloning Stem Cells 2002;4:389–96.

28. Humpherys D, Eggan K, Akutsu H, et al. Abnormal gene expression in cloned mice derived from embryonic stem cell and cumulus cell nuclei. Proc Natl Acad Sci U S A 2002;99:12889–94.

29. Eggan K, Akutsu H, Hochedlinger K, Rideout W III, Yanagimachi R, Jaenisch R. X-chromosome inactivation in cloned mouse embryos. Science 2000;290:1578–81.

30. Betts D, Bordignon V, Hill J, et al. Reprogramming of telomerase activity and rebuilding of telomere length in cloned cattle. Proc Natl Acad Sci U S A 2001; 98:1077–82.

31. Lanza RP, Cibelli JB, Blackwell C, et al. Extension of cell life-span and telomere length in animals cloned from senescent somatic cells. Science 2000;288:665–9.

32. Tian XC, Xu J, Yang X. Normal telomere lengths found in cloned cattle. Nat Genet 2000;26:272–3.

33. Wakayama T, Shinkai Y, Tamashiro KL, et al. Cloning of mice to six generations. Nature 2000;407:318–9.

34. Wells DN, Misica PM, Tervit HR. Production of cloned calves following nuclear transfer with cultured adult mural granulosa cells. Biol Reprod 1999;60:996–1005.

35. Gurdon JB, Laskey RA, Reeves OR. The developmental capacity of nuclei transplanted from keratinized skin cells of adult frogs. J Embryol Exp Morpho 1975;34:93–112.

36. Wilmut I, Schnieke AE, McWhir J, Kind AJ, Campbell KH. Viable offspring derived from fetal and adult mammalian cells. Nature 1997;385:810–3. [Erratum, Nature 1997;386:200.]

37. Liu L. Cloning efficiency and differentiation. Nat Biotechnol 2001;19:406.

38. Oback B, Wells D. Donor cells for nuclear cloning: many are called, but few are chosen. Cloning Stem Cells 2002;4:147–68.

39. Weissman IL. Stem cells: units of development, units of regeneration, and units in evolution. Cell 2000;100:157–68.

40. Vogelstein B, Albetts B, Shine K. Genetics: please don't call it cloning! Science 2002; 295:1237.

41. Colman A, Kind A. Therapeutic cloning: concepts and practicalities. Trends Biotechnol 2000;18:192–6.

42. Zwaka TP, Thomson JA. Homologous recombination in human embryonic stem cells. Nat Biotechnol 2003;21:319–21.

43. Schuldiner M, Yanuka O, Itskovitz-Eldor J, Melton DA, Benvenisty N. Effects of eight growth factors on the differentiation of cells derived from human embryonic stem cells. Proc Natl Acad Sci U S A 2000;97:11307–12.

44. Odorico JS, Kaufinan DS, Thomson JA. Multilineage differentiation from human embryonic stem cell lines. Stem Cells 2001;19:193–204.

45. Carpenter MK, Inokuma MS, Denham J, Mujtaba T, Chiu CP, Rao MS. Enrichment of neurons and neural precursors from human embryonic stem cells. Exp Neurol 2001;172: 383–97.

46. Reubinoff BE, Itsykson P, Turetsky T, et al. Neural progenitors from human embryonic stem cells. Nat Biotechnol 2001;19:1134–40.

47. Schuldiner M, Eiges R, Eden A, et al. Induced neuronal differentiation of human embryonic stem cells. Brain Res 2001;913:201–5.

48. Kaufman DS, Hanson ET, Lewis RL, Auerbach R, Thomson JA. Hematopoietic colony-forming cells derived from human embryonic stem cells. Proc Natl Acad Sci U S A 2001;98:10716–21.

49. Assady S, Maor G, Amit M, Itskovitz-Eldor J, Skorecki KL, Tzukeman M. Insulin production by human embryonic stem cells. Diabetes 2001;50:1691–7.

50. Wichterle H, Lieberam I, Porter JA, Jessell TM. Directed differentiation of embryonic stem cells into motor neurons. Cell 2002;110:385–97.

51. Kim JH, Auerbach JM, Rodriguez-Gomez JA, et al. Dopamine neurons derived from embryonic stem cells function in an animal model of Parkinson's disease. Nature 2002;418:50–6.

52. Soria B, Roche E, Bema G, Leon-Quinto T, Reig JA, Martin F. Insulin-secreting cells derived from embryonic stem cells normalize glycemia in streptozotocin-induced diabetic mice. Diabetes 2000;49:157–62.

53. Kyba M, Perlingeiro RC, Daley GQ. HoxB4 confers definitive lymphoid-myeloid engraftment potential on embryonic stem cell and yolk sac hematopoietic progenitors. Cell 2002;109:29–37.

54. Klug MG, Soonpaa MH, Koh GY, Field LJ. Genetically selected cardiomyocytes from differentiating embryonic stem cells form stable intracardiac grafts. J Clin Invest 1996;98:216–24.

55. Li M, Pevny L, Lovell-Badge R, Smith A. Generation of purified neural precursors from embryonic stem cells by lineage selection. Curr Biol 1998;8:971–4.

56. Rideout WM III, Hochedlinger K, Kyba M, Daley GQ, Jaenisch R. Correction of a genetic defect by nuclear transplantation and combined cell and gene therapy. Cell 2002;109:17–27.

57. Buehr M, Nichols J, Stenhouse F, et al. Rapid loss of Oct-4 and pluripotency in cultured rodent blastocysts and derivative cell lines. Biol Reprod 2003;68:222–9.

58. Reik W, Maher ER. Imprinting in clusters: lessons from Beckwith-Wiedemann syndrome. Trends Genet 1997;13:330–4.

59. Moorehead RA, Sanchez OH, Baldwin RM, Khokha R. Transgenic overexpression of IGF-II induces spontaneous lung tumors: a model for human lung adenocarcinoma. Oncogene 2003;22:853–7.

60. Byrne JA, Simonsson S, Gurdon JB. From intestine to muscle: nuclear reprogramming through defective cloned embryos. Proc Nad Acad Sci U S A 2002; 99:6059–63.

61. Joshi CY, Enver T. Plasticity revisited. Curr Opin Cell Biol 2002;14:749–55.

62. Ying QL, Nichols J, Evans EP, Smith AG. Changing potency by spontaneous fusion. Nature 2002;416:545–8.

63. Terada N, Hamazaki T, Oka M, et al. Bone marrow cells adopt the phenotype of other cells by spontaneous cell fusion. Nature 2002;416:542–5.

64. Vassilopoulos G, Wang P-R, Russell DW. Transplanted bone marrow regenerates liver by cell fusion. Nature 2003;422:901–4.

65. Wang X, Willenbring H, Akkari Y, et al. Cell fusion is the principal source of bone-marrow-derived hepatocytes. Nature 2003;422:897–901.

66. Morshead CM, Benveniste P, Iscove NN, van der Kooy D. Hematopoietic competence is a rare property of neural stem cells that may depend on genetic and epigenetic alterations. Nat Med 2002;8:268–73.

67. Wagers AJ, Sherwood RI, Christensen JL, Weissman IL. Little evidence for developmental plasticity of adult hematopoietic stem cells. Science 2002;297:2256–9.

68. Check E. Second cancer case halts gene-therapy trials. Nature 2003;421:305.

69. Jiang Y, Jahagirdar BN, Reinhardt RI, et al. Pluripotency of mesenchymal stem cells derived from adult marrow. Nature 2002;418:41–9.

70. Niwa H, Miyazaki J, Smith AG. Quantitative expression of Oct-3/4 defines differentiation, dedifferentiation or self-renewal of ES cells. Nat Genet 2000;24:372–6.

71. Nichols J, Zevnik B, Anastassiadis K, et al. Formation of pluripotent stem cells in the mammalian embryo depends on the POU transcription factor Oct4. Cell 1998;95:379–91.

72. Hübner K, Fuhrmann G, Christenson LK, et al. Derivation of oocytes from mouse embryonic stem cells. Science 2003;300:1251–6.

Index

abortion, 6, 35, 38–39, 45, 46, 54. *See also* Supreme Court, United States
achondroplasia, 39–40, 43–44, 45. *See also* genetic counseling
American Academy of Pediatrics, 100, 107
American College of Medical Genetics, 100, 103, 106
Amish, 41–42
Appiah, K. Anthony, 125–26, 127
Aristotle, 6–7
Asch, Adrienne, 36, 45–46, 49, 55n2
assisted reproduction: Alzheimer disease (AD) and, 76–77, 78; artificial insemination, 77, 117, 118; California Cryobank, 119, 122, 124, 125; HIV-positive women, 77–78; in vitro fertilization (IVF), 1, 2, 75, 77, 92, 115, 117, 118, 128n5, 149, 150, 153, 154, 177; limits on, 76, 77, 78; parent-child relationship and, 63, 68; postmenopausal, 76, 77; preimplantation diagnosis and, 4, 75, 76, 78; race-specific gamete selection, 115–16, 121, 122, 123–24, 125, 126, 129n19; racial attitudes involved in, 119–20, 124–25, 127, 129n19; racial disparities, 117, 118;

sex selection, 76; sperm donation, 63, 68, 77, 119–27 passim, 129n19; surrogate mothers, 2, 128n3. *See also* infertility; reproductive rights

Bérubé, Michael, 46; *Life As We Know It*, 47–48
biomedical research: public funding of, 147–48, 158–59, 166
biotechnology, 1; government regulation, 5, 6, 9; patents, 131; political implications, 3, 5, 8–9. *See also* DNA; genetic engineering; patent law
blindness, 52–54
Bush, George W.: administration policy on stem cell research, 151–67 passim, 169n12; human cloning, ban on, 168nvi

Centers for Disease Control and Prevention, 100, 104–5, 128
chimeras. *See* species identity
Clinton, Bill: administration policy on stem cell research, 150–54 passim
cloning, 5, 6, 22, 92; abnormalities related to, 173–74, *175*, 177, 178, 184, *185*, 186, 187; adult stem cells

193

and, *175*, 179, 186; blastocysts, use
of, *180*, 184, *185*, 186;
developmental totipotency, 178–79;
embryonic stem cells and, 12, 153,
173–81 passim, 184; epigenetic
reprogramming, 174, *176*, 178, *185*;
hematopoietic precursors,
differentiation of, *181*, 182, *183*;
mice, therapeutic cloning experiment
involving, 182–84; mice clones,
developmental characteristics, *175*,
177, 178, *180*, 184–*85;* murine
embryonic stem cells and, *181*, 182,
187; myeloid cells, differentiation of,
183; oocytes, use of, 176, 179, 187;
postzygotic reprogramming, 177;
prezygotic reprogramming, 176–77;
reproductive, 173, 177–78, 179–*80*,
187; somatic cells, differentiation
involving, 182, 187; therapeutic, 173,
174, 179, *180*, 181, 182, 185, 187
Congress, United States: Dickey
Amendment, 150, 151, 154, 155,
161, 166, 168nix, 169n8; human
fetal and embryo research, federal
funding for, 148–49, 150, 166

Davis, Dena: "Genetic Dilemmas and
the Child's Right to an Open
Future," 40–41, 43
deafness: as culture, 39–40; as
disability, 40–41, 42–43, 45. *See also*
genetic counseling
Department of Health, Education, and
Welfare, United States, 148, 149
Department of Health and Human
Services, United States, 107, 150, 151
Derricotte, Toi, 120–21, 122
DHEW. *See* Department of Health,
Education, and Welfare, United
States
DHHS. *See* Department of Health and
Human Services, United States
Dickey Amendment. *See* Congress,
United States

disabilities: biases against, 34, 39, 49,
50, 51, 55n1; disability rights
advocates, 33, 34, 35, 36, 55n2;
familial experiences of, 40, 43–44,
47–48, 52–54; medical views of, 45,
46, 47; social construction of, 39, 43,
49–51
DNA, 33; information, computer-
readable, 135, 136, 137, 140, 141,
144n22; methylation, 174;
mitochondrial, 22, 28n17; nuclear,
22, 28n17; patentability, 131–41
passim, 142n3, 142n7, 144n18;
recombinant, 13, 132, 133;
variability, 16. *See also* genomes;
Patent and Trademark Office, United
States
Douglas, Mary, 21
Down syndrome, 47–49. *See also*
genetic counseling

evolution, 3, 14, 16, 17, 28n11

family law, 59, 60
Federal Circuit: patent case decisions,
132, 136–37, 138, 140–41, 143n13,
144n18
Feinberg, Joel, 41, 55n4
Finkler, Kaja, 63

genetic counseling: achondroplasia,
39–40, 44–45; deafness, 39–40;
Down syndrome, 33, 46, 47, 48;
education programs, 45, 51–52;
eugenics and, 34, 37, 38;
nondirectiveness principle, 34, 36,
37, 40, 44, 45, 51; prenatal testing,
33–34, 35, 36, 38, 46, 47; standpoint
epistemology, 34–37, 43. *See also*
disabilities; population screening
genetic engineering, 2, 6, 27, 28n12. *See
also* cloning; DNA; patent law;
species identity
genetic paternity testing: constraints
proposed for, 60, 69, 70–71; disease

screening and, 60, 70; divorce and, 61, 62; fathers and, 61, 65–66, 68; laws governing, 61, 69–70; legal cases, typology of, 60–62, 65, 69; marital presumption doctrine, 59, 69–70. *See also* parenthood
genomes: *Haemophilus influenzae*, 135–36, 142n9; human, 15–16, 26, 33; Human Genome Project, 16, 19, 33, 131; Human Genome Sciences, 135, 136, 142n9
Gilligan, Carol, 35, 46–47
Gooding-Williams, Robert, 122, 123

Harding, Sandra, 36–37
human nature, 3, 5, 7, 17, 64; revolutionary political movements and, 7–8
Huxley, Aldous: *Brave New World*, 1, 2–3, 4
hybrids. *See* species identity

infertility: African American women, 115, 128n2; United States, 117, 128n8
information technology, 1–2, 8; biotechnology, compared to, 5–6
IVF. *See* assisted reproduction

Jefferson, Thomas, 4–5, 61, 67
Jews: Ashkenazi, 97, 100, 102, 103; Hassidic, 41–42, 55n3

Kass, Leon, 20; "Implications of the Human Right to Life," 37
Kent, Deborah: "Somewhere a Mockingbird," 52–54
Kittay, Eva Feder, 54

leukemia, 82, 84–85, 88, 89, 92. *See also* organ donation
liberal democracy, 3, 8

McBride, Gail, 82, 88
Mills, Charles, 120

National Commission for the Protection of Human Subjects of Biomedical and Behavioral Research, 148–49, 167nii
National Institutes of Health, 103, 148; stem cell research and, 149, 153, 161, 163, 164, 165, 167, 168nxi, 171n27
Nelson, James Lindemann, 68
NIH. *See* National Institutes of Health
nuclear technology: government regulation, 3, 6

organ donation: best interest legal standard, 84, 85–86, 87, 90–91, 92, 93n7, 93n9, 94n22, 95n32; bone marrow, 82, 84–85, 90, 94n16, 95n32; crisis in, 81; "having a child to save a child" cases, 81–92 passim; informed consent, 86, 94n23; kidney, 83–84, 85–86, 90, 91, 92n5, 93n6, 95n32; legal cases involving, 83–87, 90, 91, 93n7, 93nn10–11; minors and living-related cases, 83, 86, 87, 88, 90, 95n34; sale of organs, 87, 91, 92n2, 160; substituted judgement doctrine, 84, 85, 93n10, 94n16
Orwell, George: *1984*, 1–2, 3

parenthood: adopted children and, 63–64, 67–68; as biological relationship, 62–63, 66, 67, 68, 69; fatherhood, 62, 63, 66–67; fathers' rights advocates, 66–67; marital presumption doctrine, 65; misattributed paternity, 60, 63, 107; as social and psychological relationship, 64–70 passim
Patent and Trademark Office, United States: DNA sequences, patent eligibility of, 131, 132, 135–41 passim; "Examination Guidelines for Computer-Implemented Inventions," 137, 143nn13–14; patent appeals

process, 142–43n11; patent information, public access to, 138

patent law: computer-implemented inventions, 136, 137–38, 143n14; intellectual property rights and, 139–40, 144n23; mathematical algorithms, 137, 138, 143n14; patent bargain, 138; patent eligibility, 134, 143–44n18; patent expiration, 144n21; unpatentable subject matter, 133. *See also* biotechnology; DNA; Federal Circuit; Patent and Trademark Office, United States; Supreme Court, United States

population screening: audiometry, 100; cancer, 97; contraception, oral, and, 106; cystic fibrosis, *99*, 100, 102–3; discrimination, insurance and employment, 107, 108; genetic counseling and, 102; hereditary hemochromatosis, 104–5; hypoglycemia, 102; hypothyroidism, 98–100; informed consent, 107; newborns, 38, 98–100, 107; phenylketonuria, 97, 98, *99*; principles of, 98, *104*, 107; sickle cell disease, 98; tandem mass spectrometry, 98, 100–102; Tay Sachs disease, 97, 102; thrombophilia, 103, 105–6

pregnancy, unplanned, 88, 89, 95n31

President's Council on Bioethics, 164, 167ni, 168nvii

PTO. See Patent and Trademark Office, United States

Pufendorf, Samuel, 63

race: African American racial identity, 117–18, 119, 123, 129nn16–17, 130n32; genetics and, 117, 118, 124; housing discrimination, 121, 122; racialism, 125–26, 127; "racial navigation," 117, 122–24, 126, 127, 128n7; racial stereotyping, 116, 122, 126, 127; racism, structural, 121, 125, 129n22; white racial identity, 117, 119, 127, 129n15. *See also* assisted reproduction; infertility

religion: child-rearing and, 42; divine creation, 19–20; God, conceptions of, 19, 20, 23; Great Chain of Being, 23

reproductive rights, 75–76, 78, 88, 121–22

Roberts, Dorothy, 115, 118, 129n18

Robertson, John, 121–22

Ruddick, William, 41–42

Saxton, Marsha, 51–52

Silvers, Anita: *Disability, Difference, Discrimination*, 49

species identity: animal rights and, 23–24, 28–29n18; bestiality and, 19–20; definitions, 13, 14–15, 26, 27–28n11; essentialism, 18, 19, 24–25, 28n11; fixed, 13, 18–19, 26; homeostatic property cluster concept of, 14–15, 18, 22; human, 14, 15–18, 21–22, 27n10; interspecies hybrids and chimeras, 4, 11, 13, 19–26 passim, 28n15, 185; personhood, moral status of, 23, 24; species boundaries, 11, 12, 18, 19, 20, 24, 26, 27n8, 28n12

sperm donation. *See* assisted reproduction

stem cells: donation, informed consent to, 151, 152, 153, 154; federal research funding, 147, 150–67 passim, 168nviii, 171n27; human embryonic stem cell research, 4, 5, 11, 12, 22, 147, 150–67 passim, 168nx, 170n22, 181; neural, 11; pluripotency, 11, 22, 187; private research, 166–67, 170–71n27. *See also* Bush, George W.; Clinton, Bill; cloning; Congress, United States; National Institutes of Health

Stout, Jeffrey, 21

Supreme Court, United States: federal funding, rulings on, 170n19; patent law decisions, 132–33, 142n5; Roe v. Wade decision, 148

Wald, N. J., 98
Walker, Ann Platt: *A Guide to Genetic Counseling*, 37

Wendell, Susan: *The Rejected Body*, 50–51
Williams, Peter: "Duties and Decency," 50

Zack, Naomi, 116, 126–27

About the Editor and Contributors

Thomas A. Shannon is professor of religion and social ethics in the Department of Humanities and Arts at Worcester Polytechnic Institute.

Mark P. Aulisio is director of the Clinical Ethics Program at MetroHealth, Case Western Reserve University.

Françoise Baylis teaches at Dalhousie University.

Geoffrey D. Block is director of Liver and Pancreas Engineering program at University of Pittsburgh School of Medicine.

Rebecca S. Eisenberg is at the University of Michigan School of Law.

Hawley Fogg-Davis is in the Department of Political Science at University of Wisconsin–Madison.

Francis Fukuyama is Bernard L. Schwartz Professor of International Political Economy at the Paul H. Nitze School of Advanced International Studies of the Johns Hopkins University.

Konrad Hochedlinger is at the Whitehead Institute for Biomedical Research.

Rudolf Jaenisch is at the Department of Biology at Massachusetts Institute of Technology.

Gregory E. Kaebnick is the editor of the *Hastings Center Report*.

Muin J. Khoury is at the Office of Genomics and Disease Prevention, Centers for Disease Control and Prevention, Atlanta.

Roberta Springer Loewy is in the Bioethics Program at the University of California, Davis.

Thomas May is director of graduate studies in bioethics at the Medical College of Wisconsin.

Linda L. McCabe and Edward R. B. McCabe are at the Department of Human Genetics and Pediatrics, the David Geffen School of Medicine at the University of California, Los Angeles.

Annette Patterson is a genetic counselor at the Center for Cancer and Blood Disorders at the University of Texas Southwestern Medical School in Dallas.

Jason Scott Robert teaches at Dalhousie University.

Martha Satz is assistant professor of English at Southern Methodist University in Dallas, Texas.

Dena Towner is in the Department of Obstetrics and Gynecology at the University of California, Davis.